Alexandre Ziegler

A Game Theory Analysis of Options

Contributions to the Theory of Financial Intermediation in Continuous Time

Springer

Author

332.645
Z 66 g

Alexandre Ziegler
56A Chemin du Château
CH-1245 Collonge-Bellerive
Switzerland

```
           Library of Congress Cataloging-in-Publication Data


Ziegler, Alexandre, 1975-
   A game theory analysis of options : contributions to the theory of
financial intermediation in continuous time / Alexandre Ziegler.
      p.   cm. -- (Lecture notes in economics and mathematical
systems ; 468)
   Rev. version of the author's thesis (doctoral--University of St.
Gallen).
   Includes bibliographical references.
   ISBN 3-540-65628-6 (softcover)
   1. Options (Finance)--Prices--Mathematical models.  2. Game
theory.   I. Title.  II. Series.
HG6024.A3Z54   1999
332.64'5--dc21                                              99-17524
                                                               CIP
```

ISSN 0075-8442
ISBN 3-540-65628-6 Springer-Verlag Berlin Heidelberg New York

© Springer-Verlag Berlin Heidelberg 1999
Printed in Germany

Typesetting: Camera ready by author
SPIN: 10699924   42/3143-543210 - Printed on acid-free paper

JK

Lecture Notes in Economics
and Mathematical Systems

468

Springer
Berlin
Heidelberg
New York
Barcelona
Hong Kong
London
Milan
Paris
Singapore
Tokyo

To my parents

Acknowledgements

This book is a revised version of my doctoral dissertation submitted to the University of St. Gallen. I would like to express my gratitude to the members of my thesis committee, Professor Heinz Zimmermann and Professor Heinz Müller, both from the University of St. Gallen, for their precious assistance and encouragement. Although heavily occupied by research and teaching, they accepted to accompany me in this fascinating journey into financial economics and provided me with useful suggestions and motivating inputs.

In addition, Professor Robert Merton, Harvard Business School, and Professor Didier Cossin, University of Lausanne, provided me with interesting literature on some aspects of this thesis.

I have also greatly benefited from insightful comments and suggestions by Christophe Lamon and Matthias Aerni, both from the University of St. Gallen.

Furthermore, I would like to thank Professor Hans Schmid, Research Institute for Labor Economics and Labor Law, University of St. Gallen, for providing me with the flexibility needed to produce this dissertation and interesting research projects at the Institute.

I am also deeply indebted to Alfonso Sousa-Poza and Professor Werner Brönnimann, both from the University of St. Gallen, for their invaluable help in correcting my English, and to Olivier Kern, University of Bern, for his willingness to analyze the formal aspects of this dissertation. All errors remain mine.

Last, but by no means least, I would like to express my gratitude to my family, colleagues and friends for providing the environment and encouragement required to complete this dissertation.

Stanford, December 1998   Alexandre Ziegler

Foreword

Modern option pricing theory was developed in the late sixties and early seventies by F. Black, R. C. Merton and M. Scholes as an analytical tool for pricing and hedging option contracts and over-the-counter warrants. However, already in the seminal paper by Black and Scholes, the applicability of the model was regarded as much broader. In the second part of their paper, the authors demonstrated that a levered firm's equity can be regarded as an option on the value of the firm, and thus can be priced by option valuation techniques. A year later, Merton showed how the default risk structure of corporate bonds can be determined by option pricing techniques. Option pricing models are now used to price virtually the full range of financial instruments and financial guarantees such as deposit insurance and collateral, and to quantify the associated risks. Over the years, option pricing has evolved from a set of specific models to a general analytical framework for analyzing the production process of financial contracts and their function in the financial intermediation process in a continuous time framework.

However, virtually no attempt has been made in the literature to integrate game theory aspects, i.e. strategic financial decisions of the agents, into the continuous time framework. This is the unique contribution of the thesis of Dr. Alexandre Ziegler. Benefiting from the analytical tractability of continuous time models and the closed form valuation models for derivatives, Dr. Ziegler shows how the option pricing framework can be applied to situations where economic agents interact strategically. He demonstrates, for example, how the valuation of junior debt and capital structure decisions are affected if shareholders follow an optimal bankruptcy strategy. Other major applications of the study include the analysis of credit contracts and collateral, bank runs and deposit insurance. The careful reader will notice that the conclusions from this analysis are extremely interesting. It is my hope that Dr. Ziegler's work stimulates further research in this exciting new field, and accelerates the interaction between microeconomics and financial economics to produce new interesting insights into the structure and the functioning of the financial system.

Heinz Zimmermann
Professor of Economics and Finance
Swiss Institute of Banking and Finance
University of St. Gallen

Table of Contents

1. Methodological Issues

1.1 Introduction

Game Theory is the study of multiperson decision problems.[1] Since its beginnings in the early 20th century and von Neumann's (1928) proof of the minimax theorem, it has developed rapidly and is now a major field of economic theory. It has become a standard part of major economics textbooks and is the main analysis instrument in some important fields of economics, such as industrial organization, corporate finance and financial intermediation. In spite of this development, in recent years, game theory has faced methodological problems in handling uncertainty and timing decisions in dynamic models. This constitutes a severe limitation for the analysis of strategic issues in financial decision-making, where uncertainty and risk are particularly important.

Option Pricing is devoted to the valuation of options and, by extension, of other contingent claims. Since the pioneering work of Black and Scholes (1973) and Merton (1973), option pricing has found its way into many domains of economics. Some examples are the pricing of corporate securities, which are essentially contingent claims on the firm's asset value, and the analysis of the value of managerial flexibility and of some timing decisions in what has become known as the real options literature. Today, option pricing and continuous-time finance have grown to an essential part of financial theory.

This book presents a method, the *game theory analysis of options*, combining these two powerful instruments of economic theory to enable or facilitate the analysis of dynamic multiperson decision problems in continuous time and under uncertainty. It also demonstrates in an exemplary fashion how the method can be used to analyze some stylized problems in the theory of financial intermediation.

The basic intuition of the method, which will be presented below, is to separate the problem of the *valuation* of payoffs from the analysis of *strategic interactions*. Whereas the former is to be handled using option pricing, the latter can be addressed by game theory. In the sequel, it is demonstrated how both instruments can be combined and how game theory can be applied to complex problems of corporate finance and financial intermediation. In this respect, the method and the examples

---

[1] See Gibbons (1992), p. xi.

presented below can be understood as an attempt to integrate game theory and option pricing. Straightforward applications of the method are:

- the *pricing* of contingent claims when strategic behavior on the part of economic agents is possible,
- the analysis of the *incentive effects* of some common contractual financial arrangements, and
- the design of *incentive contracts* aiming at resolving conflicts of interest between the economic agents.

Before presenting the method in detail and turning to the examples of the following chapters, a few basic concepts of game theory and option pricing shall be introduced.

1.2 Game Theory Basics: Backward Induction and Subgame Perfection

Consider the game pictured in Figure 1.1. In period 1, player I chooses either strategy U or strategy D. In period 2, player II chooses either strategy L or strategy R. In period 3, payoffs are received, where the vector $(x; y)$ states that player I receives x and player II receives y. For example, if player I chooses U and player II chooses L, then player I will receive 2 and player II gets 1. It is assumed that all the above is *common knowledge*, that is, known to both of the players.

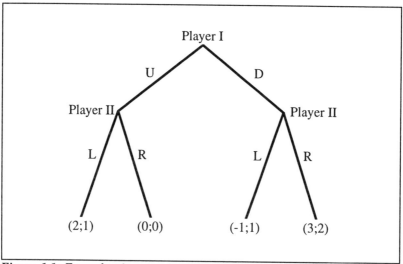

Figure 1.1: Example of a game tree (Source: Fudenberg and Tirole (1991), p. 85).

Which strategies are the players going to choose? To answer this question, the principle of *backward induction* can be used. It says that the game should be solved beginning with the *last* decision to be made, which is to be replaced by its optimal value. That is, one first solves for player II's optimal decision in the last stage, substitutes the resulting payoffs in the game tree and then works backward to find player I's optimal choice. At the node on the left, player II gets a payoff of 1 if he chooses L, and 0 if he chooses R. Therefore, he will choose L. Similarly, at the node on the right, his optimal strategy is to choose R, since this enables him to get a payoff of 2 instead of 1. Substituting these values into the game tree yields the result depicted in Figure 1.2.

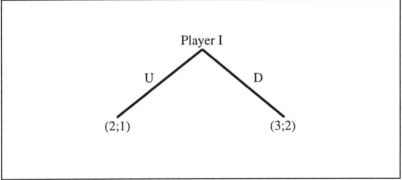

Player I

U   D

(2;1)   (3;2)

Figure 1.2: The game of Figure 1.1 after the subgames have been replaced with the optimal decisions in the last stage.

It is now straightforward to find the solution to this game: in period 1, player I can choose between a payoff of 2 (strategy U) and a payoff of 3 (strategy D). Hence, he will choose D. Actually, substituting player II's optimal strategies into the game before solving for player I's optimal strategy means that, when making his choice, player I *anticipates* player II's subsequent optimal choice.

As defined above, backward induction can only be applied to games of *perfect information*, in which all the information sets are singletons. In a game of perfect information, players move one at a time and each player knows all previous moves when making his decision.[2] This was obviously the case of the game described above where player I first made his choice between U and D, and then player II chose between L and R, *knowing* what player I had chosen. Now consider the game depicted in Figure 1.3. The subgame on the right is a simultaneous-move game. At the time he

---

[2] See Fudenberg and Tirole (1991), p. 80.

has to make his decision, player II does *not* know what player I chooses. Hence, backward induction cannot be used to determine player I's optimal choice. The idea of backward induction can, however, be extended to handle these kind of games.

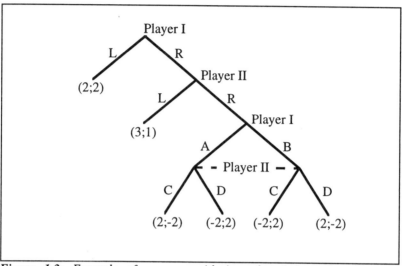

Figure 1.3: *Example of a game with imperfect information (Source: Fudenberg and Tirole (1991), p. 94).*

To see why, consider the subgame on the right. If it is reached, then each player will choose each strategy with probability ½, thus yielding an expected payoff of 0.[3] The subgame can therefore be replaced with its equilibrium payoff $(0; 0)$, yielding the game in Figure 1.4, which can then be solved using backward induction as was done with the game of Figure 1.1. To do so, note that if the node on the right is reached, Player II will choose L, thus obtaining a payoff of 1 instead of 0 if he chooses R. When making his own choice in the first stage, Player I will anticipate Player II's subsequent choice and the resulting payoff of 3. Since this is greater than the payoff of 2 he would get by playing L, he will choose R. In equilibrium, Player I chooses R and Player II L; the simultaneous-move subgame on the right is not reached.

The idea that each subgame should be replaced with its equilibrium payoff is called *subgame perfection*. Note that in a finite game of perfect information, backward induction and subgame perfection are equivalent.[4]

---

[3] This is called a mixed strategy. See Fudenberg and Tirole (1991), p. 5.
[4] See Fudenberg and Tirole (1991), p. 96.

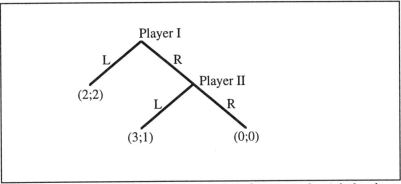

Figure 1.4: *The game of Figure 1.3 after the subgame on the right has been replaced with its equilibrium payoff.*

1.3 Option Pricing Basics: The General Contingent Claim Equation

Consider a contingent claim that is written on an underlying asset S whose value follows a geometric Brownian motion

$$dS = \mu Sdt + \sigma Sd\tilde{z} ,\qquad(1)$$

where μ is the *drift* and σ the instantaneous *standard deviation* of the process and $d\tilde{z}$ denotes the increment of a standard Wiener process. Throughout this book, it will be assumed that asset values follow such a process.[5]
Let S denote the current value of the underlying asset, t time, r the risk-free rate of return, a the payout to the holders of the underlying asset per unit time, and b be the payout to the holders of the contingent claim per unit time. Let $F(S,t)$ denote the value of the contingent claim. Then, as Merton (1977) has shown, F must satisfy the following linear partial differential equation:

$$\tfrac{1}{2}\sigma^2 S^2 F_{SS} + (rS - a)F_S + F_t - rF + b = 0 ,\qquad(2)$$

where subscripts to F denote partial derivatives. At this point, it is important to stress that the value of *any* contingent claim written on S must satisfy (2); different contingent claims only differ by their boundary conditions.
Equation (2) contains a certain number of *parameters*, and is subject to *boundary conditions*. These factors will provide the basis for the analyses conducted in this text. The game theory analysis of options presented below is concerned with the economic agents' incentives to influence the

---

[5] Huang (1985, 1987) supplies sufficient conditions under which equilibrium prices follow such processes.

parameters or the *boundary conditions* of options embedded in their economic activities.

1.4 The Method of Game Theory Analysis of Options

The method of game theory analysis of options is an attempt to combine game theory and option pricing. Using option pricing, arbitrage-free values for the payoffs to the economic agents can be obtained. These values are then inserted into the strategic games between the agents, which can thus be analyzed more realistically.

The essence of the method can be summarized as a three-step procedure:

- *First*, the game between the players is defined, that is, the players' action sets, the sequence of their choices and the resulting payoffs are specified.
- *Second*, the players' future uncertain payoffs are valued using option pricing theory. All the players' possible actions enter the valuation formula as *parameters*.
- *Finally*, starting with the last period, the game is solved for the players' optimal strategies using backward induction or subgame perfection.

In effect, the game theory analysis of options replaces the maximization of *expected utility* encountered in classical game theory models with the maximization of the value of an *option*, which gives the arbitrage-free value of the payoff to the player and can therefore be considered as a proxy for expected utility. Over the expected-utility approach, the option-pricing approach has the advantage that it automatically takes the time value of money and the price of *risk* into account.

The greatest strength of the method, however, lies in its separating the valuation problem (Step 2) from the analysis of the strategic interaction between the players (Step 3). This feature is very useful in the analysis, because complex decision problems under uncertainty can be solved by applying classical optimization procedures (minimization and maximization) to the value of the option. The analysis then often boils down to finding a first-order condition for a maximum or minimum.

To better understand how the method works, suppose that the structure of the game is the following: First, player I chooses a strategy A. Once this choice is made, player II chooses a strategy B. These strategies, together with the *future* value of the state variable S, determine the payoffs to each of the players. Let $G(A,B,S)$ and $H(A,B,S)$ denote the *current* arbitrage-free value of the payoffs to player I and II as given by option pricing, respectively. As mentioned earlier, this value is obtained by solving a differential equation similar to (2) subject to appropriate boundary conditions. The players' strategies consist in choosing one of the parameters of this differential equation or its boundary conditions so as to maximize the value of their payoffs.

In the last stage of the game, player II chooses that strategy B which maximizes the value of his expected payoff $H(A,B,S)$, that is, sets

$$\frac{\partial H(A,B,S)}{\partial B} = 0, \tag{3}$$

provided that B is not a boundary solution. This first-order condition can be solved to yield an optimal strategy $\overline{B} = \overline{B}(A,S)$, which might depend on player I's strategy choice A. Now, at the time he makes his decision, player I must anticipate player II's subsequent choice. That is, he sets

$$\frac{dG(A,\overline{B},S)}{dA} = \frac{\partial G(A,\overline{B},S)}{\partial A} + \frac{\partial G(A,\overline{B},S)}{\partial B}\frac{d\overline{B}}{dA} = 0, \tag{4}$$

thus yielding an optimal strategy $\overline{A} = \overline{A}(S)$. The term

$$\frac{\partial G(A,\overline{B},S)}{\partial B}\frac{d\overline{B}}{dA} \tag{5}$$

in expression (4) reflects the indirect effect of player I's strategy choice on his expected payoff that results from the influence of his choice on player II's optimal strategy $\overline{B}$. It captures the essence of backward induction, i.e. that player I must *anticipate* what player II will do when making his choice.

1.5 An Example: Determining the Price of a Perpetual Put Option

Consider a financial intermediary active in a competitive market and selling a perpetual put option on an underlying asset S with an exercise price of X to an investor. Which price should he ask for? To answer this question, the simple method presented above is applied.

1.5.1 Step 1: Structure of the Game

The structure of the game, which is depicted in Figure 1.5 below, is the following: At initial time, the intermediary sells the option to the investor for a certain price P_∞. The investor then holds the option until he decides to exercise it, where $\overline{S}$ denotes his optimal exercise strategy. At the time of exercise, the payoff to the investor equals the (positive) difference between the strike price X and the current value of the underlying asset $\overline{S}$, $Max[0; X - \overline{S}]$.

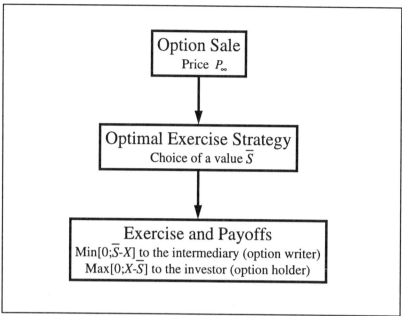

Figure 1.5: Structure of the option pricing game. In the first phase, the intermediary sells a perpetual put option to the investor at a price P_∞. The investor then chooses his optimal exercise strategy $\overline{S}$. Finally, if the investor chooses to exercise the option, he receives $X - \overline{S}$ from the intermediary.

1.5.2 Step 2: Valuing the Option

The second step in the method is to determine the arbitrage-free value of the perpetual put option, $P_\infty(S)$, *given* the option holder's exercise strategy $\overline{S}$. This value is given by the following ordinary differential equation:[6]

$$\tfrac{1}{2}\sigma^2 S^2 P_\infty'' + rSP_\infty' - rP_\infty = 0,\qquad(6)$$

subject to the boundary conditions

$$P_\infty(\infty) = 0,\qquad(7)$$

$$P_\infty(\overline{S}) = X - \overline{S}.\qquad(8)$$

The general solution to (6) is

$$P_\infty(S) = \alpha_1 S + \alpha_2 S^{-\gamma},\qquad(9)$$

where

---

[6] The analysis of this section follows the line of Merton (1973).

$$\gamma \equiv \frac{2r}{\sigma^2}. \tag{10}$$

From boundary condition (7), we have $\alpha_1 = 0$. On the other hand, boundary condition (8) requires that

$$P_\infty(\overline{S}) = X - \overline{S} = \alpha_2 \overline{S}^{-\gamma} \iff \alpha_2 = (X - \overline{S})\overline{S}^\gamma. \tag{11}$$

Hence, the value of the perpetual put option for a given exercise strategy $\overline{S}$ is given by

$$P_\infty(S) = (X - \overline{S})\overline{S}^\gamma S^{-\gamma} = (X - \overline{S})\left(\frac{S}{\overline{S}}\right)^{-\gamma}. \tag{12}$$

1.5.3 Step 3: Solving the Game

At this point, the investor's exercise strategy $\overline{S}$ is still unknown. It is a so-called *free boundary*. There are basically two approaches to compute it.

1.5.3.1 Smooth Pasting

The first method is to require P_∞ to satisfy the so-called smooth-pasting condition

$$\left.\frac{\partial P_\infty(S)}{\partial S}\right|_{S=\overline{S}} = \frac{d\overline{P}_\infty}{d\overline{S}}, \tag{13}$$

where $\overline{P}_\infty$ denotes the value of the put option upon exercise as specified by boundary condition (8): $\overline{P}_\infty = X - \overline{S}$. Using this methodology, the optimal exercise strategy can be computed by setting

$$\left.\frac{\partial P_\infty(S)}{\partial S}\right|_{S=\overline{S}} = \left.\frac{\partial}{\partial S}\left((X - \overline{S})\left(\frac{S}{\overline{S}}\right)^{-\gamma}\right)\right|_{S=\overline{S}} = \left.\left(-\frac{\gamma}{S}(X - \overline{S})\left(\frac{S}{\overline{S}}\right)^{-\gamma}\right)\right|_{S=\overline{S}} \tag{14}$$

$$= -\frac{\gamma}{\overline{S}}(X - \overline{S}) = \frac{d\overline{P}_\infty}{d\overline{S}} = \frac{d}{d\overline{S}}(X - \overline{S}) = -1,$$

thus yielding

$$\frac{\gamma}{\overline{S}}(X - \overline{S}) = 1 \iff \overline{S} = \frac{\gamma}{1+\gamma}X. \tag{15}$$

1.5.3.2 Value-Maximizing Exercise Strategy

An alternative way of finding the free boundary $\overline{S}$ is to require that it *maximize* the value of the option, that is, that the option holder only exercises when it is *optimal* to do so, thus setting

$$\frac{\partial P_\infty}{\partial \overline{S}} = -\left(\frac{S}{\overline{S}}\right)^{-\gamma} + \frac{2r}{\sigma^2}(X - \overline{S})\left(\frac{S}{\overline{S}}\right)^{-\gamma} \cdot \frac{1}{\overline{S}} = 0 \qquad (16)$$

and obtaining

$$\overline{S} = +\gamma(X - \overline{S}) \iff \overline{S}(1+\gamma) = \gamma X . \qquad (17)$$

Thus, the underlying asset value $\overline{S}$ under which exercise is optimal is given by

$$\overline{S} = \frac{\gamma}{1+\gamma} X , \qquad (18)$$

which is the same solution as that given by smooth-pasting.

1.5.3.3 Link between the two Approaches

A question that obviously arises is that of the link between the two approaches. Are they related? Do they always yield similar results? Merton (1973) showed that smooth-pasting is actually implied by value-maximization: Let $f(x;\overline{x})$ be a differentiable function for $0 \le x \le \overline{x}$ and let $\partial^2 f / \partial \overline{x}^2 < 0$. Set $h(\overline{x}) = f(\overline{x};\overline{x})$, where h is a differentiable function of $\overline{x}$.

Let $\overline{x} = \overline{x}*$ be the $\overline{x}$ which maximizes f, i.e.

$$\left.\frac{\partial f(x;\overline{x})}{\partial x}\right|_{\overline{x}=\overline{x}*} = 0. \qquad (19)$$

Now consider the total derivative of f with respect to $\overline{x}$ along the boundary $\overline{x} = \overline{x}*$:

$$df = \left.\frac{\partial f}{\partial x}\right|_{x=\overline{x}} d\overline{x} + \frac{\partial f}{\partial \overline{x}} d\overline{x} . \qquad (20)$$

From the definition of h,

$$\frac{dh}{d\overline{x}} = \left.\frac{\partial f}{\partial x}\right|_{x=\overline{x}} + \frac{\partial f}{\partial \overline{x}} . \qquad (21)$$

Now, at the point $\overline{x} = \overline{x}*$, $\partial f / \partial \overline{x} = 0$ by (19), so

$$\frac{dh}{d\overline{x}} = \left.\frac{\partial f}{\partial x}\right|_{x=\overline{x}} , \qquad (22)$$

which is the smooth-pasting condition (13).[7]

---

[7] This result is known as the "envelope theorem" in microeconomics. See Simon and Blume (1994), pp. 452-457.

1.5.3.4 Alternative Behavioral Assumptions

What is, then, the difference between the two approaches? It clearly lies in the different behavioral assumptions on the part of the economic agents. Whereas smooth pasting is more a mathematical property (tangency), value maximization has a clear intuitive economic basis consisting in the "optimal" behavior of economic agents. In the remainder of this book, value maximization shall therefore be used, thus stressing that the problems analyzed are multiperson games in which the players behave optimally.

1.5.4 The Solution

The solution to our problem of finding which price the intermediary should ask for the option can now be found as follows. The option writer can be expected to anticipate the investor to exercise when is optimal to do so. Therefore, he would ask for a price equal to the value of the option if the holder exercises optimally, that is, a price equal to the value given by (12) with the exercise strategy given by (15) and (18), thus yielding

$$P_\infty(S) = (X - \bar{S})\left(\frac{S}{\bar{S}}\right)^{-\gamma} = \frac{X}{1+\gamma}\left(\frac{(1+\gamma)S}{\gamma X}\right)^{-\gamma}, \qquad (23)$$

which can therefore be expected to be the market price of the perpetual put option. This solution was already derived by Merton (1973), who implicitly used the method described above.

1.6 Outline of the Book

The following chapters illustrate how the game theory analysis of options can be applied to some classical problems of corporate finance and financial intermediation. While the examples provided in the sequel are of great interest as such, the methodological emphasis of this book is equally important.

Chapter 2, *Credit and Collateral*, analyzes two classical problems of financial contracting, namely, the risk-shifting problem and the observability problem, and shows that they are very closely related. More specifically, it demonstrates that – except in the special case of full collateralization – there exists no contract solving both the risk-shifting and the observability problem simultaneously and discusses the practical implications of this result for corporate financing.

Chapter 3, *Endogenous Bankruptcy and Capital Structure*, develops a model of the firm with outside (debt) financing and endogenous bankruptcy. In analyzing the last stage of the game, namely, the shareholders' bankruptcy decision, it is shown that endogenous bankruptcy gives rise to a principal-agent problem. The resulting agency cost of debt is measured. Then, the firm's investment decision is addressed. Underinvestment and risk-shifting are studied and the role and incentive effects of debt covenants discussed. Subsequently, optimal capital structure and its properties are described. It is shown that optimal capital structure depends on the interest rates charged on the loan given to the firm. Some properties of the ex ante optimal capital structure are discussed. Moving back through the game, equilibrium on the credit market is then presented. Finally, an incentive contract allowing the lender to lead the borrower to declare bankruptcy at a pre-specified asset value is constructed.

Chapter 4, *Junior Debt*, is devoted to the incentive effects of subordinated debt. Extending the model of Chapter 3 to the case where there are many lenders of different seniority, the analysis presented illustrates how the existence of junior debt influences the borrower's bankruptcy decision. Then, his incentives to issue junior claims are discussed and it is demonstrated that such an issue may result in a wealth transfer between security holders, thus leading to a distortion in the borrower's incentive to issue junior debt. Consequences for the firm's capital structure are explored.

Chapter 5, *Bank Runs,* analyzes this important phenomenon and its incentive effects. After exploring the depositors' decision to run on a bank, the bank's equity is valued under the run restriction. The decision of the bank's shareholders to recapitalize the bank is analyzed. Finally, the bank's optimal investment choice when bank runs are possible is explored and the consequences for the funding of banks are discussed.

Chapter 6, *Deposit Insurance,* discusses the costs and benefits of deposit insurance and its incentive effects. It demonstrates that deposit insurance does *not* result in risk-shifting behavior on the part of banks if the guarantor is perfectly informed and can seize the assets immediately. Some interesting incentive problems might arise, however, if the guarantor cannot observe current asset value or has to wait before he can seize the assets. Possible incentive contracts between the bank and the guarantor aiming at addressing these issues are presented and discussed.

Chapter 7, *Summary and Conclusions*, summarizes the main results of the book and discusses the strengths and weaknesses of the game theory analysis of options. While the method presented here allows a better analysis of strategic interactions under uncertainty in dynamic settings, it is subject to some severe limitations, namely, mathematical complexity and the fact that continuous-time finance is a mere approximation of reality. By replacing the maximization of *expected utility* encountered in classical game theory models with the maximization of the value of an *option*, the game theory analysis of options allows to solve complex decision problems under uncertainty by applying classical optimization procedures (minimization and maximization) to the value of the option. Over the expected-utility approach, the option-pricing approach has the advantage that it automatically takes the time value of money and the price of *risk* into account. Its greatest strength, however, lies in its ability to separate *valuation* from the analysis of *strategic behavior* – a feature that isn't displayed by classical game theory models, where difficulties in valuing uncertainty complicate the analysis of strategic interactions.

2. Credit and Collateral

2.1 Introduction

Moral hazard is a widespread source of inefficiency in economics. In financial contracting, both classical forms of moral hazard exist, each giving rise to specific incentive issues.[1] In a situation of *hidden action*, the agent takes an action that is not observed by the principal. For example, the borrower might try to influence the return distribution of his project to increase his expected payoff at the expense of the lender. This is the so-called *risk-shifting* or asset substitution problem, which was first laid out by Jensen and Meckling (1976). In contrast, in a situation of *hidden information*, the agent privately observes the true state of the world prior to choosing an observable action. In the context of financial contracting, the borrower typically is the only person that can observe project returns at no cost. To the extent that his promised payment depends positively on realized project return, he might have an incentive to understate project return in order to reduce his payment to the lender. This form of information asymmetry gives rise to the so-called *observability* problem, which was addressed in the costly state verification literature in the wave of Townsend's (1979) pathbreaking paper. The main conclusion of this literature is that costly state verification by the principal (lender) makes complete risk-sharing suboptimal.

While these two strands of literature each provide interesting insights into the optimal structure of financial contracts, they have not been properly integrated. The aim of this chapter is to analyze the risk-shifting and the observability problem using the instruments provided by the game theory analysis of options and to demonstrate how they are related. The setting used is voluntarily simple, with a given contract life and a single terminal payment from the borrower to the lender; more complicated situations involving interim payments and timing issues will be addressed in subsequent chapters.

The structure of the chapter is as follows: Section 2.2 analyzes the *risk-shifting* problem using a simple principal-agent framework in which the principal lends money to the agent for a finite period of time and cannot call the loan back before term. Extending the basic intuition of early models that convexity in the agent's payoff is responsible for risk-shifting, a contract avoiding risk-shifting is developed. It is shown that, in

---

[1] See Müller (1997), p. 2 for a general introduction to moral hazard.

continuous-time analysis, there exists an infinity of contracts having this property. However, only one of these contracts is able to solve the risk-shifting incentive of the agent at any point in time and for any value taken by the state variable. The concept of a *dynamically stable* (or renegociation-proof) incentive contract is introduced. This contract is expected to be preferred by both principal and agent because it avoids costly renegociation. The optimal contract is a linear risk-sharing contract. This linearity result has the interesting intuitive interpretation of leading the lender to buy equity and generalizes previous results in the literature on the optimality of linear contracts. It confirms Lemma 1 of Seward (1990), which states that if the firm's return is observable, then the appropriate investment incentives can be restored through the use of an all-equity financial structure.

Turning to the *observability* problem, Section 2.3 shows that the analysis of one-period models still holds. Because output cannot be observed by the principal, the contractual payment cannot be made contingent upon it. Therefore, the optimal contract when output is unobservable is a debt contract. Under such a contract, however, the agent has an incentive to engage in risk-shifting behavior. This incentive problem can be mitigated through the use of collateral.

Finally, Section 2.4 concludes the chapter and presents practical consequences of the general result that there exists no contract solving *both* the risk-shifting and the observability problem simultaneously.

2.2 The Risk-Shifting Problem

A classical problem in financial contracting is the so-called risk-shifting problem. This term stands for the incentive the borrower has to influence the risk of his project in order to increase the value of his payoff at the expense of the lender.

Consider a financial intermediary that provides a firm with capital for investment, and assume that the intermediary knows that the firm has an incentive to increase the risk of its project. There are three basic approaches to solve this problem. The first is for the intermediary to simply anticipate the behavior of the agent and ask for a higher interest rate on his loan. The second approach calls for the intermediary to closely monitor the agent to avoid his taking undue risk. Finally, the intermediary can try to design a contract to have the agent behave properly without having to monitor him. Because of the resulting savings in interest and monitoring costs, this last approach to solving the risk-shifting problem can be expected to prevail in practice.[2] In this section, we therefore analyze the structure of the risk-shifting problem using the game theory analysis of options and show how an incentive contract can be designed that ensures that the agent does not engage in undue risk.

2.2.1 The Model

Consider a financial intermediary, the principal, that lends money to an agent for investment in one of many projects that are available only to the agent. Assume that the principal cannot observe the project choice of the agent, and therefore cannot assess the risk of the project. At initial time, all projects have the same price S_0, but different risks. The project value then evolves according to a geometric Brownian motion. Assume, further, that the agent can, at any time, change his mind and switch to another project at no cost.[3] More specifically, the agent can choose to invest in a series of projects whose dynamics are given by

$$dS_i = \mu_i S_i dt + \sigma_i S_i d\tilde{z}_i .$$ (1)

For simplicity, assume that all projects have a finite life of T and a random terminal value $\overline{S}_i$ observable by both the principal and the agent. Assume that the principal and the agent agree on a single, end-of-period contingent payment to the principal $f(\overline{S}_i)$.[4] Suppose that the principal and the agent have no other assets and limited liability. Then, the *effective* payoff to the principal, whatever has been agreed upon, is given by

$$Min\left[\overline{S}_i ; f\left(\overline{S}_i\right)\right].$$ (2)

The payoff to the agent equals the difference between total project return and the amount paid out to the principal:

$$\overline{S}_i - Min\left[\overline{S}_i ; f\left(\overline{S}_i\right)\right] = Max\left[0; \overline{S}_i - f\left(\overline{S}_i\right)\right].$$ (3)

Figure 2.1 summarizes the structure of this game. In the first stage, the financing contract is signed. Then, the agent invests in a project and may, if he wishes, switch to a project involving more or less risk. Finally, at

---

[2] As shown in Stiglitz and Weiss (1981), raising the interest rate might be unprofitable for banks because of the resulting adverse selection effects.
[3] Note that this implies that there are no scale effects, i.e. the amount invested in the new project can always be chosen to equal the proceeds from liquidating the old project. Alternatively, one could think of the model as involving a single project, but with many alternative business strategies of different riskiness.
[4] See Chapter 3 for the analysis of a debt contract involving interim interest payments and Chapter 4 for the case of several debt contracts.

expiration of the contract, the return on the project is observed by both principal and agent and the agent pays $Min[\overline{S}_i; f(\overline{S}_i)]$ to the lender.

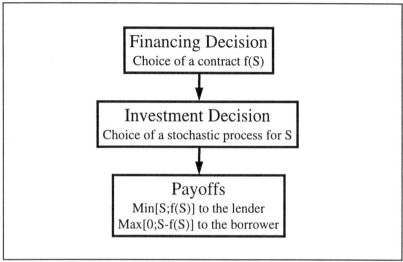

Figure 2.1: *Structure of the game between lender and borrower. After the financing contract is signed, the borrower chooses an investment project. At any time during the life of the contract, he can switch to another project involving higher or lower risk. At time T, project return is publicly observed and payoffs are received.*

2.2.2 Valuing the Players' Payoffs

From these assumptions, we can easily see that a contract basically is a payment by the principal today with an agreement by the agent to pay him $Min[\overline{S}_i; f(\overline{S}_i)]$ at time T. The sum of the payoff to the agent and the principal at T is $\overline{S}_i$. Within this feasibility constraint, the principal and the agent can agree on any payment. An example of such a payment scheme is depicted in Figure 2.2.

Any contract between the principal and the agent can be characterized by a fixed payment D and a certain number of put and call options, as is demonstrated in Figure 2.3.

2.2.3 Developing an Incentive Contract

The structure of a contract of the form described in Figure 2.3 that avoids strategic risk-taking or risk-avoidance by the agent shall now be determined. By assumption, the value of the payment to the principal can be calculated as

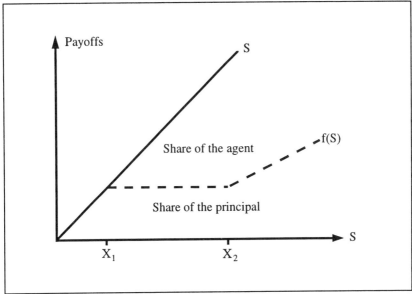

Figure 2.2: *Example of a feasible profit-sharing rule between the principal and the agent. The rule attributes everything to the principal up to an amount X_1, plus the half of any amount in excess of X_2.*

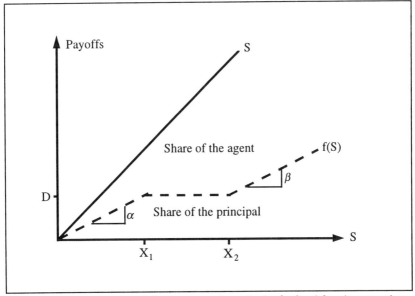

Figure 2.3: *A payment of the agent to the principal of α (short) put options with an exercise price of X_1, β call options with an exercise price X_2 and a lump sum of D.*

$$\Pi = D + \alpha \cdot P(X_1) + \beta \cdot C(X_2), \tag{4}$$

where P and C stand for the Black-Scholes put and call option values with an exercise price of X_1 and X_2, respectively:

$$P(X_1) = X_1 e^{-r\tau}\left(1 - N\left(d_1 - \sigma\sqrt{\tau}\right)\right) - S\left(1 - N(d_1)\right)$$

$$C(X_2) = SN(d_2) - X_2 e^{-r\tau} N\left(d_2 - \sigma\sqrt{\tau}\right), \tag{5}$$

where

$$d_i = \frac{\ln(S/X_i) + \left(r + \sigma^2/2\right)\tau}{\sigma\sqrt{\tau}}, \quad i \in \{1, 2\}, \tag{6}$$

$N(\cdot)$ denotes the cumulative standard normal distribution function, τ the remaining life of the loan and r the risk-free interest rate.

In order to avoid risk-shifting, the contract parameters α, β, D, X_1 and X_2 have to be chosen so that the agent has no incentive to influence the risk of the project. This can be achieved by making the arbitrage-free value of the borrower's payoff Π independent of asset risk σ. Formally, we must have

$$\frac{\partial \Pi}{\partial \sigma} = \alpha \frac{\partial P(X_1)}{\partial \sigma} + \beta \frac{\partial C(X_2)}{\partial \sigma} = 0. \tag{7}$$

From option pricing theory, we know that[5]

$$\frac{\partial P(X_1)}{\partial \sigma} = \frac{S\sqrt{\tau}}{\sqrt{2\pi}} e^{-d_1^2/2},$$

$$\frac{\partial C(X_2)}{\partial \sigma} = \frac{S\sqrt{\tau}}{\sqrt{2\pi}} e^{-d_2^2/2}. \tag{8}$$

Hence, the incentive compatibility condition (7) becomes

$$\frac{\partial \Pi}{\partial \sigma} = \alpha \frac{\partial P(X_1)}{\partial \sigma} + \beta \frac{\partial C(X_2)}{\partial \sigma}$$

$$= \frac{S\sqrt{\tau}}{\sqrt{2\pi}}\left(\alpha e^{-\frac{\left(\ln(S/X_1) + \left(r + \sigma^2/2\right)\tau\right)^2}{2\sigma^2\tau}} + \beta e^{-\frac{\left(\ln(S/X_2) + \left(r + \sigma^2/2\right)\tau\right)^2}{2\sigma^2\tau}} \right) = 0. \tag{9}$$

Examination of equation (9) shows that, with α, β, X_1 and X_2 free, there exists an *infinity* of incentive compatible profit-sharing contracts. Which

---

[5] See Hull (1993), p. 315.

one should the principal and the agent choose? To answer this question, let us introduce the concept of a dynamically stable incentive contract.

2.2.4 Dynamically Stable (Renegociation-Proof) Incentive Contracts

Definition: An incentive contract is *dynamically stable (renegociation-proof)* if it assures proper incentives at any point in time over the life of the contract and for any value of the state variable S.[6]

This concept is intuitively appealing. If incentive compatibility is not satisfied as time passes or when the value of the state variable changes, then the principal and the agent can gain mutually by *renegociating* the contract. To the extent that renegociation involves costs, they will be able to gain if they can agree on a contract that assures proper incentives over its whole life. We should therefore expect dynamically stable incentive contracts to prevail in practice. Figure 2.4 gives an example of a contract that is *not* renegociation-proof: for low asset values, the borrower has an incentive to increase project risk in order to lower the value of the payment to the lender. In other words, for low asset values, $\partial \Pi / \partial \sigma < 0$, and the risk-shifting incentive is given by $-\partial \Pi / \partial \sigma > 0$. For high asset values, the borrower can reduce the value of the lender's claim by lowering project risk since $\partial \Pi / \partial \sigma > 0$.

Proposition 1: The only dynamically stable contract of type (4) is a contract for which $X_1 = X_2$ and $\alpha + \beta = 0$.

Proof: For the optimality condition (9)

$$\frac{\partial \Pi}{\partial \sigma} = \frac{S\sqrt{\tau}}{\sqrt{2\pi}} \left(\alpha e^{-\frac{\left(\ln(S/X_1)+\left(r+\sigma^2/2\right)\tau\right)^2}{2\sigma^2 \tau}} + \beta e^{-\frac{\left(\ln(S/X_2)+\left(r+\sigma^2/2\right)\tau\right)^2}{2\sigma^2 \tau}} \right) = 0$$

to hold for any value of τ and S, we must have

$$\alpha e^{-\frac{\left(\ln(S/X_1)+\left(r+\sigma^2/2\right)\tau\right)^2}{2\sigma^2 \tau}} + \beta e^{-\frac{\left(\ln(S/X_2)+\left(r+\sigma^2/2\right)\tau\right)^2}{2\sigma^2 \tau}} = 0 \quad \forall S, \tau. \tag{10}$$

This condition can be rewritten as

---

[6] The classical moral hazard literature uses the term renegociation-proofness to describe a contract that is never revised. See Müller (1997), p. 13. In this chapter, the term *dynamical stability* is used to stress that renegociation would exclusively be triggered by a change in the value of the state variable or the passage of time.

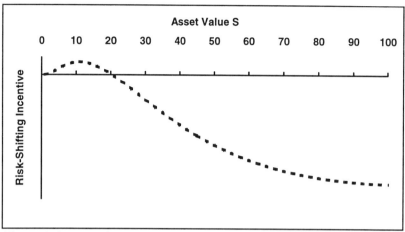

Figure 2.4: *Example of a non-renegociation-proof contract with a positive risk-shifting incentive for low project values and a negative risk-shifting incentive for high project values. (Parameter values: $\alpha = -0.5$, $\beta = 0.5$,*
$X_1 = 25$, $X_2 = 50$, $r = 0.05$, $\sigma = \tau = 1$.)

$$-\frac{\alpha}{\beta} = e^{\frac{\left(\ln(S/X_1)+\left(r+\sigma^2/2\right)\tau\right)^2 - \left(\ln(S/X_2)+\left(r+\sigma^2/2\right)\tau\right)^2}{2\sigma^2\tau}} \qquad \forall S, \tau. \qquad (11)$$

For a contract to be dynamically stable, α and β must be constant. Therefore, the above expression must not depend on S and τ. Hence, we must have

$$\Psi = \frac{\left(\ln(S/X_1)+\left(r+\sigma^2/2\right)\tau\right)^2 - \left(\ln(S/X_2)+\left(r+\sigma^2/2\right)\tau\right)^2}{2\sigma^2\tau}$$

$$= \frac{\ln(X_2/X_1)\left(2\ln(S)-\ln(X_1X_2)+2\left(r+\sigma^2/2\right)\tau\right)}{2\sigma^2\tau} \qquad (12)$$

constant with respect to S and τ. Taking the partial derivative of the above expression with respect to S and setting it equal to zero gives

$$\frac{\partial\Psi}{\partial S} = \frac{\ln(X_2/X_1)}{S\cdot\sigma^2\tau} = 0, \qquad (13)$$

that is

$$X_1 = X_2. \qquad (14)$$

Substituting this condition back into (11) gives

$$-\frac{\alpha}{\beta} = e^0 = 1, \qquad (15)$$

and therefore
$$\alpha + \beta = 0 , \tag{16}$$
the desired result.

2.2.5 The Feasible Dynamically Stable Incentive Contract

With the conditions $X_1 = X_2$ and $\alpha + \beta = 0$, the profit-sharing rule agreed upon by the principal and the agent to save renegociation costs is *linear* in $\overline{S}$. The value of the constant payment D and of the parameters α and β has now to be determined. This can be easily done using the feasibility condition (2).

Because the agent cannot pay out more than $\overline{S}$ to the principal, α must be negative. To see this, suppose α were chosen to be positive. In this case, the contract would call for the agent to make a positive payment to the principal when the project ends worthless, that is, when $\overline{S}$ is zero, which violates the feasibility constraint (2). So α must be negative and β positive, so that the variable component of the payment to the principal is given by $f(\overline{S}) = \beta\overline{S}$, where β is a positive constant.

Consider now the fixed payment D. To be feasible, the contract must call for a fixed payment D of zero. To see this, assume first that D were chosen to be positive. Then, the agent could not fulfill his contractual obligation whenever $\overline{S} < D + \beta\overline{S}$, i.e. whenever $\overline{S} < D / (1 - \beta)$, which would create a risk-shifting problem. Similarly, if D were chosen to be negative, a risk-shifting problem would arise as well. These results can be summarized in the following proposition:

Proposition 2: The only feasible, dynamically stable incentive contract is linear in $\overline{S}$ and calls for no fixed payment by the agent. That is, the contract is given by
$$f(\overline{S}) = \beta\overline{S} , \tag{17}$$
where β is a positive constant.

The result in Proposition 2 has a simple intuitive interpretation: When terminal project value is perfectly observable, there is no reason for the lender to ask for a fixed payment, because this would only impede risk-sharing and create risk-incentive issues without providing any benefits. Therefore, the lender agrees to receive a *proportional* share of β in the firm's gross return, i.e. buys *equity*. This confirms Lemma 1 in Seward (1990): If the firm's return is completely observable, then the appropriate investment incentives can be restored through the use of an *all-equity* financial structure.

2.2.6 The Financing Decision

With Proposition 2, which gives the structure of the feasible dynamically-stable incentive contract, it is now possible to determine how much the lender will be ready to give to the borrower at initial time. If the lender lends an amount D_0 to receive a share β of the terminal payoff $\overline{S}$, then he will at most agree to lend

$$D_0 = \beta S_0, \tag{18}$$

where S_0 denotes the total initial investment in the project. Accordingly, the borrower, which receives a share $1 - \beta$ of the terminal payout, must provide a share $1 - \beta$ in equity capital.

2.2.7 The Effect of Payouts

An interesting question that arises in the context of project financing is that of how the analysis has to be modified if the borrower receives payouts from the project before maturity. Intuitively, one would expect the lender to ask for more equity capital in this case, since the terminal payoff to the lender is reduced by the amount of payouts. To show that this is indeed the case, consider the simple case in which the borrower can withdraw a continuous proportional dividend of δ from the project. One can show that the value of the project without this right to dividends is given by[7]

$$S_{ex} = Se^{-\delta T}, \tag{19}$$

where T is the life of the project. This effect is depicted in Figure 2.5. Using (19), the no-expected-loss condition (18) becomes:

$$D_0 = \beta S_0 e^{-\delta T}, \tag{20}$$

and hence the lender will accept to finance at most a share of

$$\beta' = \frac{D_0}{S_0} = \beta e^{-\delta T} \tag{21}$$

of the project for a right to a share β of the terminal payoff. This result has an important implication: since the lender can at most receive the whole terminal payoff (i.e. β is bounded above by 1 because of limited liability), some projects with high payout rates or a long life might not be feasible. To see this, set $\beta = 1$ in equation (21). Then,

---

[7] See Ingersoll (1987), p. 367 f.

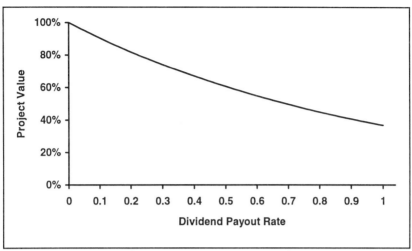

Figure 2.5: *Current value (as a percentage of initial investment) of a one-year project without the right to dividends as a function of the dividend payout rate δ. As the dividend payout rate increases, the value of the claim on the project without the right to dividends is reduced.*

$$\beta' = \frac{D_0}{S_0} = e^{-\delta T}, \tag{22}$$

and the borrower must provide at least an amount

$$E_0 = (1 - \beta')S_0 = (1 - e^{-\delta T})S_0 \tag{23}$$

in equity capital. To the extent that he does not have this amount available, the project cannot be realized. There might therefore be welfare costs to premature payouts.

One could of course argue that lender and borrower could reach an agreement to preclude early payouts by the agent. While such restrictions might work for pecuniary payouts, they cannot address the problem of fringe benefits.

2.3 The Observability Problem

The analysis above assumed that both principal and agent could observe the terminal value of the investment at no cost. This assumption, however, is not realistic. In many situations, the agent can be expected to be better informed about the success of his project than the principal. Our analysis has to take into account the possibility of the agent lying to the principal when reporting realized project return. Under the incentive contract derived above, the agent has a strong incentive to understate the true success of the project, and as a result of this misinformation, having to pay less to the principal.

To see this, assume both parties have agreed on the above contract, calling for the agent to pay a share β of the gross return of the project to the principal. Clearly, if the principal cannot verify the true return of the project, the agent will save β dollars for each dollar he understates the return. Therefore, the return announcement by the agent has no interior optimum. The best strategy for the agent is to announce a zero gross return and pay nothing to the principal. However, in this case, the principal can be expected to *anticipate* the behavior of the agent and lend him no money. Clearly, this would lead the agent to forgo profitable investment opportunities, leading to a socially suboptimal outcome. Both principal and agent therefore have an incentive to find a solution to this problem.

2.3.1 Costly State Verification

Townsend (1979) analyzes the problem of costly state verification in a one-period setting. Townsend assumes that the project return has a continuous, strictly positive density function $g(S)$ in an interval $[\alpha, \beta]$, $\alpha > 0$, and that lender and borrower can agree in advance as to when verification should take place or not. He then shows that, when only *pure* verification strategies are allowed, the optimal contract has the following properties:[8]

- the payment to the lender is equal to some constant amount D whenever verification does not take place,
- the verification region is a lower interval $[\alpha, \gamma)$, $\gamma \le \beta$, that is, verification will occur whenever the announced project payoff S is lower than γ.

This contract has properties that are very similar to those of a standard debt contract, in which a fixed payment D is specified and verification occurs whenever bankruptcy is declared, that is, when $S < D$. Thus, costly state verification makes complete risk-sharing suboptimal.

Interpreting state verification as bankruptcy, Gale and Hellwig (1985) show that the optimal (debt) contract, by leading to a maximal repayment in bankruptcy states, allows the fixed repayment in non-bankruptcy states to be minimized, thus minimizing the probability of bankruptcy and hence the costs.

---

[8] Townsend also shows, however, that, in a discrete state-space, this pure verification agreement can be dominated by a *stochastic* verification procedure in which the lender only verifies with a probability $\xi < 1$ if a bad state is announced.

The result that the promised payoff to the principal should be *constant* whenever verification does not take place implies that there exists no contract that solves *both* the risk-shifting and the observability problem simultaneously. To see why, remember that the only contract that avoids risk-shifting is such that the principal receives a constant *share* of realized project returns. Such a contract, however, can only be compatible with the above solution to the observability problem if verification *always* occurs. But in this case, verification costs are maximized, which is clearly suboptimal.

Risk-shifting incentives of debt contracts are endemic to the convex structure of the payoff to the borrower that results from a constant payment in good states, i.e. when terminal project return is high. As the subsequent analysis demonstrates, these adverse incentives can be mitigated through the use of collateral.

2.3.2 Collateral

Suppose that lender and borrower come to the agreement that the borrower is to provide the lender with collateral in amount X. Assume that the contract is a standard debt contract, which calls for payment of a fixed amount D at maturity, that the loan is not fully collateralized (that is, that $X < D$) and that the life of the loan is fixed at T (early repayment is therefore precluded). Then, the payoff to the lender at maturity is

$$Min[D; X + S - c],\tag{24}$$

where c denotes the fixed verification cost borne by the lender if bankruptcy occurs (Figure 2.6).

The payoff to the borrower is given by

$$Max[0; X + S - D] = Max[0; S - (D - X)].\tag{25}$$

This payoff structure is the same as that of a call option on S with exercise price $D - X$, as Figure 2.7 illustrates.

Hence, the expected payoff to the borrower is equal to the value of this call option:

$$C = S \cdot N(d) - (D - X)e^{-r\tau} N(d - \sigma\sqrt{\tau}),\tag{26}$$

where

$$d = \frac{\ln\left(\dfrac{S}{D - X}\right) + \left(r + \dfrac{1}{2}\sigma^2\right)\tau}{\sigma\sqrt{\tau}},\tag{27}$$

$N(\cdot)$ denotes the cumulative standard normal distribution function and τ the remaining life of the loan.

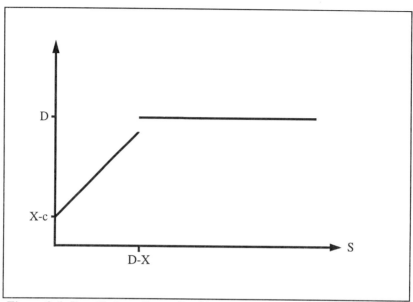

Figure 2.6: Payoff of a collateralized loan to the lender.

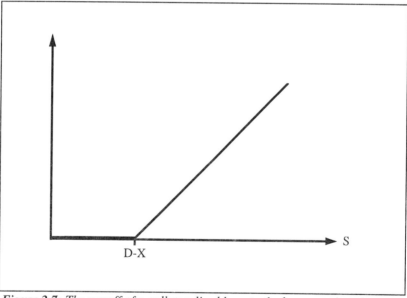

Figure 2.7: The payoff of a collateralized loan to the borrower has the same structure as that of a call option on S with an exercise price $D - X$.

To explore the influence of the existence of collateral on the agent's risk-shifting incentives, compute the partial derivative of his expected payoff with respect to σ.

$$\frac{\partial C}{\partial \sigma} = \frac{S\sqrt{\tau}}{\sqrt{2\pi}} e^{-d^2/2}. \tag{28}$$

This expression is positive, so that the risk-shifting problem exists. It is, however, mitigated as the amount of collateral increases. If $X \uparrow D$, that is, if the loan is almost fully collateralized, we have

$$\lim_{X \uparrow D} d = \lim_{X \uparrow D} \frac{\ln\left(\frac{S}{D-X}\right) + \left(r + \frac{1}{2}\sigma^2\right)\tau}{\sigma\sqrt{\tau}}$$

$$= \frac{\ln(S) + \left(r + \frac{1}{2}\sigma^2\right)\tau}{\sigma\sqrt{\tau}} - \frac{\lim_{X \uparrow D} \ln(D-X)}{\sigma\sqrt{\tau}} = +\infty \tag{29}$$

and therefore

$$\lim_{X \uparrow D} \frac{\partial C}{\partial \sigma} = \lim_{d \downarrow -\infty} \frac{S\sqrt{\tau}}{\sqrt{2\pi}} e^{-d^2/2} = 0. \tag{30}$$

Only in the limiting case where the loan is fully collateralized does the risk-shifting problem disappear. The reason for this result is that when the lender's claim is fully secured by collateral, it becomes totally riskless. Since the value of the lender's claim does not depend on project risk anymore, the borrower is unable to reduce it by shifting project risk. In this respect, collateral can be understood as a contractual device influencing the borrower's risk-shifting incentive.

Figure 2.8 illustrates this fact by plotting the risk-shifting incentive (28) for different collateral amounts. It demonstrates that the amount of collateral guaranteeing the loan has a dramatic influence on the risk-shifting incentive of the borrower. As the amount of collateral is increased, the risk-shifting incentive increases for low project values and falls for high project values. In the limiting case of full collateralization, the risk-shifting incentive disappears completely.

Collateral therefore protects the lender in *two* distinct ways: first, it grants him a claim on an additional asset in the case of bankruptcy, thus allowing him to recover more wealth. Second, and somewhat less obviously, it mitigates the borrower's incentives to shift risk, thus reducing the probability of the bankruptcy (verification) region being reached.[9] If the loan is fully secured by collateral, the borrower's risk-shifting incentive disappears and its interests become aligned with those of the lender.

---

[9] This dual function of contractual devices will also appear in the case of loan covenants (see Section 3.5.4 below).

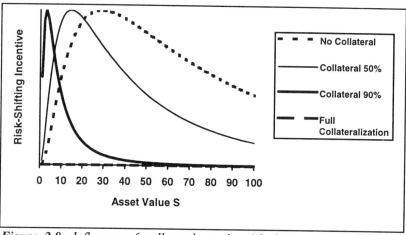

Figure 2.8: Influence of collateral on the risk-shifting incentive of the borrower for the following parameter values: $D = 50$, $r = 0.05$, $\sigma = \tau = 1$. As the amount of collateral is increased, the risk-shifting incentive increases for low project values and falls for high project values. In the limiting case of full collateralization, the risk-shifting incentive disappears completely.

2.4 Conclusion

This chapter used the game theory analysis of options to address two classical problems in financial contracting, the risk-shifting problem and the observability problem, and to explore the relationship between them.

The term risk-shifting stands for the borrower's incentive to influence the risk of his project in order to increase the value of his payoff at the expense of the lender. The analysis presented in Section 2.2 showed that, at any point in time, there exists an infinity of profit-sharing rules avoiding risk-shifting. However, only one of these contracts was found to be dynamically stable (renegociation-proof): the linear profit-sharing contract, in which the lender receives a proportional share of the project's gross return.

The observability problem arises because the lender is unable to observe project return at no cost. If a linear profit-sharing scheme is agreed upon in order to avoid risk-shifting, then the borrower has an incentive to understate the project return. The reason is that in so doing, he reduces the amount due to the lender. To address the observability problem, a fixed payment can therefore be agreed upon. The contract agreed upon by lender and borrower then takes the form of a standard debt contract.

The analysis presented in Section 2.3 demonstrated that solving the observability problem through a (concave) standard debt contract creates a risk-shifting problem, and that any attempt to solve the risk-shifting problem through a linear risk-sharing contract creates a revelation problem. Hence, except in the case of fully collateralized debt, which is effectively risk-free, both problems cannot be solved simultaneously. The reason is that risk-taking incentives of less-than-fully-collateralized loan contracts are endemic to the existence of limited liability and the associated convex structure of the payoff to the borrower (see John et al. (1991)).

This fact has an important practical implication. Suppose the lender wishes to finance two companies, one having no opportunity for risk-shifting but a value that is costly to monitor (say, a huge industrial corporation) and the other having risk-shifting opportunities but observable returns (say, a startup). Then, he should finance the industrial firm with *debt* and buy *equity* in the startup. While this rule may be useful in deciding which contractual arrangement to choose when one incentive problem clearly dominates the other, it is of no help if both the risk-shifting and the observability problem are acute. In other words, there exists a fundamental trade-off between solving both problems.

3. Endogenous Bankruptcy and Capital Structure

3.1 Introduction

This chapter uses the game theory analysis of options to analyze the *principal-agent* problem created by endogenous bankruptcy, its implications for the firm's capital structure choice, investment decision and the structure of debt contracts. Moreover, the effects of loan covenants on the borrower's incentives to increase asset risk and the payout rate are analyzed, and it is demonstrated that when bankruptcy is endogenous, monitoring asset value and monitoring asset risk can be considered as substitutes.

The analysis presented in this chapter is closely related to the work of Leland (1994) and Chesney and Gibson (1994). Leland (1994) analyzes the problem of endogenous bankruptcy. While conceptually drawing on his model, this chapter provides an analysis that differs from that of Leland (1994) in several respects. First, the model presented below considers two distinct types of interests on a loan, an effective payment and an increase in the face value of debt. In doing so, different incentive effects of these two components of debt service can be analyzed. Second, the analysis below discusses endogenous bankruptcy as a principal-agent problem and quantifies the agency cost of debt. Third, stressing the game theory perspective of this book and furthering the analysis of Chapter 2, the effect of loan covenants and of the lender's information about asset value on risk-taking incentives is presented. Fourth, some interesting properties of optimal capital structure are explored. Finally, possible incentive contracts aiming at achieving specific goals are developed.

It is interesting to contrast some of the results in this chapter with those of Chesney and Gibson (1994). Modeling firm equity as a knock-out call option, Chesney and Gibson (1994) analyze the risk incentive effects of debt. While their approach is quite similar to that taken in the sequel, two important differences deserve mention. First, they use a finite, given firm life of T, whereas we use an *infinite* horizon (that is, a perpetual down-and-out option). The rationale for doing so is that in practice, firms are not closed or liquidated as their debt matures. Rather, as long as it pays to do so, their life is extended through the issue of new debt. Therefore, the appropriate model for the Chesney and Gibson (1994) approach would be that of a knock-out option with extendible maturity, which could be priced according to the method developed by Longstaff (1990). The approach taken in this chapter does this implicitly. By specifying a continuous interest payment with default on this payment triggering immediate liquidation, the analysis below de facto models firm equity as a knock-out option with an instantaneously extendible maturity, where the instantaneous interest payment is the premium to be paid in exchange for an infinitesimal extension of the option's maturity.

The approach presented below differs from that presented in Chesney and Gibson (1994) in a second respect, however. Whereas they use a fixed, exogenous knock-out boundary, the analysis in this chapter treats bankruptcy as *endogenous*. This alternative specification yields results that are quite different from those presented in Chesney and Gibson (1994). While they are able to derive an interior optimal firm risk, the analysis below demonstrates that, absent covenants, equity holders would always wish to increase firm risk. The reason is that, in doing so, they lower the knock-out boundary, thus raising the value of equity.

3.2 The Model

Consider a lender (say, a financial intermediary) and a borrower (say, a firm) that reach the following agreement: in exchange for a loan of F, the borrower is to pay an instantaneous interest of $\phi D(t)dt$ to the lender, where $D(t) = D_0 e^{r^* t}$ denotes the face value of debt and ϕ is the instantaneous interest rate to be effectively paid on debt. Asset sales are prohibited. Hence, any net cash outflows associated with interest payments must be financed by selling additional equity. This setting generalizes the model in Leland (1994) insofar as it distinguishes between effective interest payments $\phi D(t)dt$ and the increase in the face value of debt, which occurs with the rate r^*. Throughout, it is assumed that $r^* < r$, where r denotes the risk-free interest rate.

Assume that the firm is liquidated if (and only if) the borrower defaults on his interest payments to the lender. If bankruptcy occurs, a fraction $0 \leq \alpha < 1$ of value is lost, leaving debt holders with $(1-\alpha)S_B$, where S_B denotes the asset value at which bankruptcy occurs.

The structure of the game between lender and borrower is summarized in Figure 3.1. In the first phase, the *financing* decision is made. The amount of debt, D, and the interest rates r^* and ϕ are determined. Once this financing is done, the firm chooses its *investment* program. Finally, equity holders choose their *bankruptcy* strategy. If the firm goes bankrupt, its assets are liquidated and payoffs are realized.

Throughout the chapter, any conflicts of interest between management and equity holders are ignored. Rather, it is assumed that the firm (or its management) makes the decisions that lie in the best interest of equity holders, possibly at the expense of creditors.

The structure of the chapter is as follows: Section 3.3 values the firm and its different securities. Section 3.4 analyzes the last stage of the game, namely, the bankruptcy decision of the equity holders. It demonstrates that the equity holders' bankruptcy choice is suboptimal from the standpoint of the lender. This conflict provides a rationale for some commonly observed characteristics of loan contracts. The agency cost resulting from the socially suboptimal bankruptcy strategy of the equity holders is measured. Using the results on bankruptcy, Section 3.5 analyzes possible problems in the firm's investment decision, namely, the underinvestment problem and the risk-shifting problem. Then, Section 3.6 derives the optimal capital structure and discusses its properties. Section 3.7 develops an incentive contract aiming at realizing a prespecified bankruptcy behavior of the borrower. Section 3.8 presents an extension of the model to a different payout setting. Section 3.9 summarizes the main results and insights of the chapter.

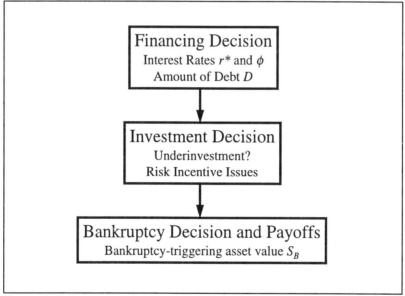

Figure 3.1: *Structure of the game. In the first phase, lender and borrower sign a contract specifying the face amount of debt, D, and the interest rates r* and φ. Then, the borrower chooses his investment strategy. Finally, if the equity holders default on interest payments, the firm is liquidated and payoffs are received.*

3.3 The Value of the Firm and its Securities

Once the game has been specified, the next step is to value the players' payoffs using option pricing theory, treating all the players' decision variables as parameters. This constitutes the object of this section. In Section 3.4, we will then begin the analysis of the players' optimal strategies with the last decision to be made, namely, the equity holders' bankruptcy decision.

3.3.1 The Value of Debt

The value of the borrower's assets, S, is assumed to follow the usual geometric Brownian motion

$$dS = \mu S dt + \sigma S d\tilde{z} \ . \tag{1}$$

For the time being, assume that asset substitution is not possible, so that the parameters μ and σ are known to the lender. Since asset sales are prohibited, any net cash outflows associated with interest payments must be financed by selling additional equity. The value of the lender's claim satisfies the following differential equation:

$$\tfrac{1}{2}\sigma^2 S^2 F_{SS} + rSF_S + r*D(t)F_{D(t)} - rF + \phi D(t) = 0 \ , \tag{2}$$

where subscripts to F denote partial derivatives. Note that F depends on t only through the face value of debt $D(t)$. Making the change in variables[1]

$$V = \frac{S_t}{D(t)} \tag{3}$$

and defining

$$G(V) = \frac{F(S)}{D(t)} , \tag{4}$$

one can show that G satisfies the following ordinary differential equation:[2]

$$\tfrac{1}{2}\sigma^2 V^2 G'' + (r - r*)VG' - (r - r*)G + \phi = 0 \ . \tag{5}$$

The general solution is

$$G = \alpha_0 + \alpha_1 V + \alpha_2 V^{-\gamma*} , \tag{6}$$

where

---

[1] See Merton (1990), p. 298.

[2] From the definitions in equations (3) and (4), $F_S = G'$, $F_{SS} = G''/D$, $F_D = G - VG'$. Substituting these expressions into (2) yields

$$\tfrac{1}{2}\sigma^2 S^2 \frac{G''}{D} + rSG' + r*D(G - VG') - rDG + \phi D = 0 \ .$$

Collecting terms and crossing out D gives (5). See Ingersoll (1987), p. 380 for an example of this method.

$$\gamma^* \equiv 2\frac{r-r^*}{\sigma^2}. \tag{7}$$

Substituting the original variables back into (6) yields the following expression for the value of the debt, F:

$$F = \alpha_0 D(t) + \alpha_1 S + \alpha_2 \left(D(t)\right)^{1+\gamma^*} S^{-\gamma^*}. \tag{8}$$

This value must satisfy the following boundary conditions:

$$F(S_B) = (1-\alpha)S_B, \tag{9}$$

$$F(\infty) = \frac{\phi D(t)}{r-r^*}. \tag{10}$$

Equation (9) stems from the properties of the bankruptcy process; equation (10) states that bankruptcy becomes irrelevant and the debt risk-free as S becomes very large. From (10), α_1 must be zero. Hence, we can write

$$F = \alpha_0 D(t) + \alpha_2 \left(D(t)\right)^{1+\gamma^*} S^{-\gamma^*}. \tag{11}$$

Then, from (10), we have

$$\alpha_0 = \frac{\phi}{r-r^*}. \tag{12}$$

Finally, using (9), we obtain

$$F(S_B) = (1-\alpha)S_B = \frac{\phi}{r-r^*}D(t) + \alpha_2 \left(D(t)\right)^{1+\gamma^*} S_B^{-\gamma^*}, \tag{13}$$

and

$$\alpha_2 = \frac{(1-\alpha)S_B - \dfrac{\phi}{r-r^*}D(t)}{\left(D(t)\right)^{1+\gamma^*} S_B^{-\gamma^*}} = \left(\frac{S_B}{D(t)}\right)^{\gamma^*}\left((1-\alpha)\frac{S_B}{D(t)} - \frac{\phi}{r-r^*}\right), \tag{14}$$

and therefore for the value of the debt, F:

$$F(S) = \frac{\phi D(t)}{r-r^*} + \left(D(t)\right)^{1+\gamma^*}\left(\frac{S_B}{D(t)}\right)^{\gamma^*}\left((1-\alpha)\frac{S_B}{D(t)} - \frac{\phi}{r-r^*}\right)S^{-\gamma^*} \tag{15}$$

$$= \frac{\phi D(t)}{r-r^*} + \left((1-\alpha)S_B - \frac{\phi D(t)}{r-r^*}\right)\left(\frac{S}{S_B}\right)^{-\gamma^*}.$$

The value of the risky debt thus equals the value of the risk-free debt, $\phi D(t)/(r-r^*)$, plus a (negative) amount that takes the expected losses in the event of bankruptcy into account. To interpret equation (15) more easily, one can rewrite it as

$$F(S) = \frac{\phi D(t)}{r - r^*}\left[1 - \left(\frac{S}{S_B}\right)^{-\gamma^*}\right] + (1 - \alpha)S_B\left(\frac{S}{S_B}\right)^{-\gamma^*}. \qquad (16)$$

Equation (16) says that the value of the risky debt equals the value of the risk-free debt, $\phi D(t)/(r - r^*)$, times the risk-neutral probability that bankruptcy does not occur, $1 - (S/S_B)^{-\gamma^*}$, plus the value of the proceeds from asset liquidation in the event of bankruptcy, $(1 - \alpha)S_B$, times the risk-neutral probability of bankruptcy, $(S/S_B)^{-\gamma^*}$.

3.3.2 The Value of the Firm

From Leland (1994), we know that the total value of the firm W reflects three terms: the firm's asset value S, the value of the tax deduction of interest payments TB, less the value of bankruptcy costs K. The value of bankruptcy costs K must satisfy (8) with boundary conditions

$$K(S_B) = \alpha S_B, \qquad (17)$$
$$K(\infty) = 0. \qquad (18)$$

From (18), $\alpha_0 = \alpha_1 = 0$, and from (17), we get

$$K(S_B) = \alpha S_B = \alpha_2\left(D(t)\right)^{1+\gamma^*}S_B^{-\gamma^*} \quad \Leftrightarrow \quad \alpha_2 = \alpha\left(\frac{S_B}{D(t)}\right)^{1+\gamma^*}, \qquad (19)$$

and therefore

$$K(S) = \alpha_2\left(D(t)\right)^{1+\gamma^*}S^{-\gamma^*} = \alpha\left(\frac{S_B}{D(t)}\right)^{1+\gamma^*}\left(D(t)\right)^{1+\gamma^*}S^{-\gamma^*}. \qquad (20)$$

$$= \alpha S_B^{1+\gamma^*}S^{-\gamma^*}$$

Similarly, the value of the tax benefits, TB, must satisfy (8) with boundary conditions

$$TB(S_B) = 0, \qquad (21)$$
$$TB(\infty) = \frac{\theta \phi D(t)}{r - r^*}. \qquad (22)$$

Boundary condition (21) states that the tax benefits are lost if bankruptcy occurs. Boundary condition (22) states that, as the asset value becomes very large and bankruptcy unlikely, the value of the tax benefits approaches the value of the risk-free debt times the tax rate θ. From (22), $\alpha_1 = 0$ and $\alpha_0 = \theta\phi/(r - r^*)$. Substituting these values into (8) and using (21) yields

$$TB(S_B) = \frac{\theta\phi D(t)}{r-r*} + \alpha_2 (D(t))^{1+\gamma*} S_B^{-\gamma*} = 0 \quad \Leftrightarrow$$

$$\alpha_2 = -\frac{\theta\phi}{r-r*}\left(\frac{S_B}{D(t)}\right)^{\gamma*}.$$

(23)

Hence,

$$TB(S) = \frac{\theta\phi D(t)}{r-r*} - \frac{\theta\phi}{r-r*}\left(\frac{S_B}{D(t)}\right)^{\gamma*} (D(t))^{1+\gamma*} S^{-\gamma*}$$

$$= \frac{\theta\phi D(t)}{r-r*}\left(1 - \left(\frac{S}{S_B}\right)^{-\gamma*}\right).$$

(24)

Using (20) and (24), the total value of the firm, W, is:
$$W(S) = S + TB(S) - K(S)$$

$$= S + \frac{\theta\phi D(t)}{r-r*}\left(1 - \left(\frac{S}{S_B}\right)^{-\gamma*}\right) - \alpha S_B^{1+\gamma*} S^{-\gamma*}.$$

(25)

Equation (25) says that the total value of the firm, W, equals current asset value, S, plus the present value of the tax shields, $\theta\phi D(t)/(r-r*)$, times the risk-neutral probability that bankruptcy does not occur, $1 - (S/S_B)^{-\gamma*}$, minus the value lost in the event of bankruptcy, αS_B, times the risk-neutral probability of bankruptcy, $(S/S_B)^{-\gamma*}$.

3.3.3 The Value of Equity

The value of equity, E, is the total value of the firm W less the value of debt F:

$$E(S) = W(S) - F(S)$$

$$= S + \frac{\theta\phi D(t)}{r-r*}\left(1 - \left(\frac{S}{S_B}\right)^{-\gamma*}\right) - \alpha S_B^{1+\gamma*} S^{-\gamma*}$$

$$- \left(\frac{\phi D(t)}{r-r*} + \left((1-\alpha)S_B - \frac{\phi D(t)}{r-r*}\right)\left(\frac{S}{S_B}\right)^{-\gamma*}\right)$$

(26)

$$= S - \frac{(1-\theta)\phi D(t)}{r-r*}\left(1 - \left(\frac{S}{S_B}\right)^{-\gamma*}\right) - S_B^{1+\gamma*} S^{-\gamma*}.$$

Equations (15), (25) and (26) give the value of debt, the firm and equity for arbitrary parameter values. Using these results, the players' optimal strategies can now be solved for, starting with the last decision of the game, namely, the equity holders' bankruptcy strategy S_B.

3.4 The Effect of Capital Structure on the Firm's Bankruptcy Decision

3.4.1 The Equity Holders' Optimal Bankruptcy Choice

The asset value which triggers bankruptcy, S_B, can now be determined. It is chosen by the equity holders so as to maximize the current value of equity, that is, by setting

$$\frac{\partial E(S)}{\partial S_B} = \frac{(1-\theta)\phi D(t)}{r-r^*}\gamma * S_B^{\gamma^*-1} S^{-\gamma^*} - (1+\gamma^*)S_B^{\gamma^*} S^{-\gamma^*} = 0, \qquad (27)$$

yielding[3]

$$S_B = \frac{(1-\theta)\phi D(t)}{r-r^*}\frac{\gamma^*}{1+\gamma^*} = \frac{(1-\theta)\phi D(t)}{r-r^*+\sigma^2/2}, \qquad (28)$$

where the second equality follows from the definition of $\gamma *$ in equation (7). Notice that this value is linear in $\phi D(t)$ and does not depend on the current asset value S. The reason for this result is the following: The product of the interest rate ϕ and the face value of debt $D(t)$ gives the premium that equity holders must pay to debt holders to keep their option alive. The higher this premium, the higher the asset value at which paying it will not be optimal from the standpoint of the equity holders and the higher therefore the bankruptcy-triggering asset value S_B.

Notice, also, that a higher asset risk σ implies a lower bankruptcy trigger S_B. The intuition for this result is clear: as asset risk rises, so does the value of equity for a given S_B. Hence, the equity holders' incentive to

---

[3] It is a maximum, since

$$\frac{\partial^2 E(S)}{\partial S_B^2} = \frac{(1-\theta)\phi D(t)}{r-r^*}\gamma *(\gamma *-1)S_B^{\gamma^*-2} S^{-\gamma^*} - (1+\gamma^*)\gamma * S_B^{\gamma^*-1} S^{-\gamma^*}$$

$$= S_B^{\gamma^*-2} S^{-\gamma^*}\gamma * \left(\frac{(1-\theta)\phi D(t)}{r-r^*}(\gamma *-1)-(1+\gamma^*)S_B\right)$$

$$= S_B^{\gamma^*-2} S^{-\gamma^*}\gamma * \left(\frac{(1-\theta)\phi D(t)}{r-r^*}(\gamma *-1)-(1+\gamma^*)\frac{(1-\theta)\phi D(t)}{r-r^*}\frac{\gamma^*}{1+\gamma^*}\right)$$

$$= -S_B^{\gamma^*-2} S^{-\gamma^*}\gamma *\frac{(1-\theta)\phi D(t)}{r-r^*} < 0.$$

default on the interest payment, i.e. on the premium to keep the option alive is weaker. This lowers the bankruptcy trigger S_B.

3.4.2 The Principal-Agent Problem of Endogenous Bankruptcy

The simple model presented above can be used to analyze the principal-agent problem of endogenous bankruptcy in loan contracts in a very general way. The incentive problem stems from the fact that borrower and lender do not agree on the appropriate asset value which is to trigger bankruptcy.

To see this, assume that the *lender* would be entitled to choose the bankruptcy-triggering point. Then, he would either set it at zero, or choose it as high as possible, that is, force bankruptcy immediately, or as soon as the condition

$$S > \frac{\phi D(t)}{(r - r^*)(1 - \alpha)} \tag{29}$$

is met.[4]

---

[4] The question to be solved here is that of the strategy S_B that maximizes the value of debt. The first-order-condition is

$$\frac{\partial F(S)}{\partial S_B} = D(t) \left[\frac{1-\alpha}{D(t)} \left(\frac{S}{S_B} \right)^{-\gamma^*} + \left((1-\alpha) \frac{S_B}{D(t)} - \frac{\phi}{r-r^*} \right) \frac{\gamma^*}{S_B} \left(\frac{S}{S_B} \right)^{-\gamma^*} \right]$$

$$= \left[(1 + \gamma^*)(1-\alpha) - \frac{\phi D(t)}{r-r^*} \frac{\gamma^*}{S_B} \right] \left(\frac{S}{S_B} \right)^{-\gamma^*} = 0.$$

Solving for S_B yields

$$S_B = \frac{\phi D(t)}{(r - r^*)(1 - \alpha)} \frac{\gamma^*}{1 + \gamma^*}.$$

However, evaluating $\partial^2 F(S) / \partial S_B^2$ shows that this point is in fact a minimum:

$$\frac{\partial^2 F(S)}{\partial S_B^2} = \left[(1 + \gamma^*)(1-\alpha) - \frac{\phi D(t)}{r-r^*} \frac{\gamma^*-1}{S_B} \right] \frac{\gamma^*}{S_B} \left(\frac{S}{S_B} \right)^{-\gamma^*}$$

$$= \left((1 + \gamma^*)(1-\alpha) - \frac{\phi D(t)}{r-r^*} \frac{\gamma^*-1}{\dfrac{\phi D(t)}{(r-r^*)(1-\alpha)} \dfrac{\gamma^*}{1+\gamma^*}} \right) \frac{\gamma^*}{S_B} \left(\frac{S}{S_B} \right)^{-\gamma^*}$$

$$= \frac{(1 + \gamma^*)(1-\alpha)}{S_B} \left(\frac{S}{S_B} \right)^{-\gamma^*} > 0.$$

But then, there are two possible optima: 0 and ∞. One can show that

But which bankruptcy trigger is socially optimal, i.e. maximizes the value of the firm as a whole? Since

$$\frac{\partial W}{\partial S_B} = -\left(\frac{\theta\phi D(t)}{r-r^*}\gamma^* S_B^{\gamma^*-1}S^{-\gamma^*} + \alpha(1+\gamma^*)S_B^{\gamma^*}S^{-\gamma^*}\right) < 0 \quad \forall S_B > 0, \quad (29)$$

the bankruptcy trigger that maximizes firm value is $S_B = 0$. These results are summarized in the following figure which plots the values of E, F and W as a function of the bankruptcy-triggering asset value S_B for different face values of debt D. As S_B rises, firm value W is reduced. Equity value E, however, rises at first, reaches a global maximum at (28) and then falls. Finally, debt value F falls at first, reaches a minimum at

$$S_B = \frac{\phi D(t)}{(r-r^*)(1-\alpha)}\frac{\gamma^*}{1+\gamma^*} \quad (30)$$

and then increases again.

The socially optimal bankruptcy trigger $S_B = 0$ is, however, not achievable. To see this, remember that the lender will liquidate as soon as $S > \phi D(t)/((r-r^*)(1-\alpha))$, and that the optimal bankruptcy trigger from the standpoint of the borrower is $S_B = (1-\theta)\phi D(t)/(r-r^*+\sigma^2/2)$. Now assume that the lender would want to enter a contract leading the borrower to choose $S_B = 0$. Then, he must set $\phi = 0$. But then we know that the value of the lender's claim equals

$$F(S_B = 0) = \frac{\phi D(t)}{r-r^*} = 0. \quad (32)$$

$$\lim_{S_B \downarrow 0} F(S) = \frac{\phi D(t)}{r-r^*}$$

and

$$\lim_{S_B \uparrow \infty} F(S) = +\infty.$$

Basically, this latter strategy should be chosen by the lender. Unfortunately, it is not feasible. A feasible strategy for the lender, however, is to liquidate if current asset value is be such that liquidating now yields more than choosing the strategy to liquidate when the asset value falls to zero, that is, if

$$F(S_B = S) = \frac{\phi D(t)}{r-r^*} + (1-\alpha)S - \frac{\phi D(t)}{r-r^*} > \frac{\phi D(t)}{r-r^*} = F(S_B = 0),$$

or

$$S > \frac{\phi D(t)}{(r-r^*)(1-\alpha)}.$$

This means that the lender will typically choose to liquidate projects doing very poorly *and* those doing very well, whichever comes first.

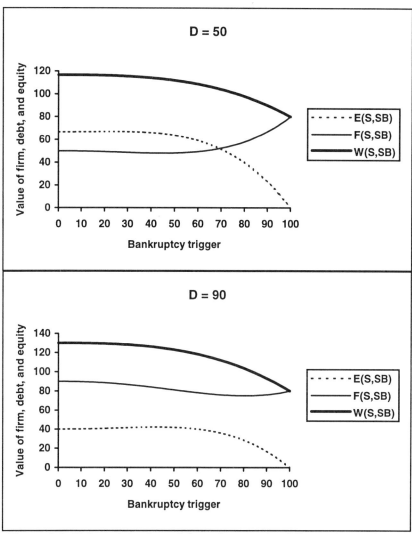

Figure 3.2: *Value of the firm, debt and equity for different face values of debt as a function of the bankruptcy-triggering asset level S_B for the following parameter values: $\theta = 1/3$, $\phi = 0.05$, $\alpha = 0.2$, $S = 100$, $r - r^* = 0.05$ and $\sigma = 0.2$. As S_B is increased, overall firm value W is reduced. Equity value E rises, and then falls. Finally, debt value F falls at first, and then rises.*

Moreover, if the lender would set $\phi = 0$, then he would wish to liquidate the position as soon as $S > 0$. Clearly, this would be suboptimal from the standpoint of the borrower.

The analysis presented here therefore provides a rationale for two very important characteristics of debt contracts, namely,

- the provision that a certain interest is to be paid *effectively,* and not merely added to the principal amount of the loan (i.e. $\phi > 0$), and
- the provision that the lender is allowed to call the loan if (and often only if) the borrower *defaults* on his interest payments.

These two characteristics arise from the fact that the lender's claim is worthless if $\phi = 0$ and from his incentive to call the loan as soon as $S > \phi D(t) / ((r - r^*)(1 - \alpha))$ if this is not prevented contractually. It is interesting to note that these two facts, which are specified exogenously in many models, arise *endogenously* in the context of this model. They are a direct consequence of the problem of choosing a bankruptcy trigger acceptable to both the lender and the borrower.

3.4.3 Measuring the Agency Cost of Debt

Using the above results on the equity holders' optimal bankruptcy decision, the agency cost of debt can be determined. By analogy with the analysis in Mello and Parsons (1992), the agency cost of debt is defined as the reduction in firm value resulting from the equity holders' choosing a socially suboptimal bankruptcy strategy. Formally, the agency cost of debt, C, equals:

$$C = W(S_B = 0) - W\left(S_B = \frac{(1-\theta)\phi D(t)}{r - r^*} \frac{\gamma^*}{1 + \gamma^*}\right). \qquad (33)$$

Using (25) yields

$$C = S^{-\gamma^*} \frac{\phi D(t)}{r - r^*}\left(\theta + \alpha(1-\theta)\frac{\gamma^*}{1 + \gamma^*}\right)\left(\frac{(1-\theta)\phi D(t)}{r - r^*}\frac{\gamma^*}{1 + \gamma^*}\right)^{\gamma^*}. \qquad (34)$$

As given by (34), the agency cost of debt C depends positively on the face value of debt $D(t)$, on the interest rate ϕ and on the bankruptcy cost α. The intuition for these results is the following: because the agency cost of debt embodies the *expected* deadweight loss resulting from bankruptcy, it will depend positively on factors leading to more frequent bankruptcy (ϕ and $D(t)$) and on the (fraction of) value lost in the event of bankruptcy, α.

3.5 The Investment Decision

Once the equity holders' optimal bankruptcy decision has been determined, their investment choices can be analyzed. More specifically, this section is concerned with two classical problems arising in the theory of corporate finance: underinvestment and risk-shifting.

3.5.1 Myers' (1977) Underinvestment Problem

Myers (1977) argues that firms may forgo profitable investment opportunities because of the existence of debt. The intuition behind this result is that by recapitalizing the firm, shareholders make debt less risky. Hence, a part of the benefits of new investment accrues to debt holders.

To analyze the underinvestment problem, we model new investment as a scaling-up of existing operations by the factor $w > 0$. That is, new investment is an impulse pushing asset value from S to $(1+w)S$. Equity value after recapitalization will equal

$$E((1+w)S) = (1+w)S - \frac{(1-\theta)\phi D(t)}{r-r*}\left(1-\left(\frac{(1+w)S}{S_B}\right)^{-\gamma*}\right)$$

$$- S_B^{1+\gamma*}\left((1+w)S\right)^{-\gamma*}. \tag{35}$$

The increase in equity value is
$$\Delta E = E((1+w)S) - E(S)$$

$$= wS - \frac{(1-\theta)\phi D(t)}{r-r*}\left(\frac{S}{S_B}\right)^{-\gamma*}\left(1-\left(1+w\right)^{-\gamma*}\right) \tag{36}$$

$$+ S_B^{1+\gamma*}S^{-\gamma*}\left(1-\left(1+w\right)^{-\gamma*}\right).$$

For additional investment to be undertaken, the increase in the value of equity must be greater than the amount invested: $\Delta E > wS$. Using (36), this condition becomes

$$S_B > \frac{(1-\theta)\phi D(t)}{r-r*}. \tag{37}$$

Substituting for the value of S_B from equation (28), condition (37) becomes

$$\frac{(1-\theta)\phi D(t)}{r-r*+\sigma^2/2} > \frac{(1-\theta)\phi D(t)}{r-r*}, \tag{38}$$

or

$$r-r*+\frac{\sigma^2}{2} < r-r* \quad \Leftrightarrow \quad \frac{\sigma^2}{2} < 0, \tag{39}$$

which is impossible. Hence, if debt cannot be renegociated, underinvestment always obtains, regardless of current asset value (Figure 3.3). The reason is that a part of the increase in overall firm value resulting from the shareholders' equity contribution accrues to debt holders.[5] Formally, the debt appreciation that would result if the new investment were undertaken equals

$$\Delta F(S) = \left((1-\alpha)S_B - \frac{\phi D(t)}{r-r^*} \right) \left(\frac{S}{S_B} \right)^{-\gamma^*} \left((1+w)^{-\gamma^*} - 1 \right) > 0 \qquad (40)$$

and is depicted in Figure 3.3.

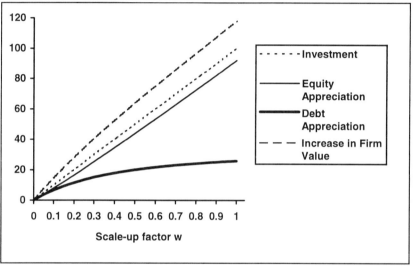

Figure 3.3: *Myers' underinvestment problem for the following parameter values:* $\theta = 1/3$, $\phi = 0.1$, $\alpha = 0.2$, $D(t) = 70$, $S = 100$, $r - r^* = 0.05$ *and* $\sigma = 0.2$. *Contributing additional equity capital is unprofitable for the shareholders because a part of the resulting increase in firm value accrues to debt holders. Hence, profitable investment is not undertaken.*

3.5.2 Risk-Shifting

The above analysis assumed that the project risk σ was known to the principal and constant. This might, however, not always be the case. The question of whether the agent has an incentive to increase risk is,

---

[5] Note that from the standpoint of overall firm value, additional investment is always beneficial, since it increases the value of the tax benefits and reduces bankruptcy costs. Formally,

$$\Delta W(S) = wS + \left(\frac{\theta \phi D(t)}{r-r^*} + \alpha S_B \right) \left(\frac{S}{S_B} \right)^{-\gamma^*} \left(1 - (1+w)^{-\gamma^*} \right) > wS.$$

therefore, of particular interest. To answer it, consider the partial derivative of the equity value (26) with respect to σ^2 :

$$\frac{\partial E(S)}{\partial \sigma^2} = \frac{\partial E(S)}{\partial \gamma^*} \frac{d\gamma^*}{d\sigma^2} = \frac{\partial}{\partial \gamma^*}\left(\left(\frac{(1-\theta)\phi D(t)}{r-r^*} - S_B\right)\left(\frac{S}{S_B}\right)^{-\gamma^*}\right) \cdot \frac{d\gamma^*}{d\sigma^2}. \quad (41)$$

Since

$$\frac{d\gamma^*}{d\sigma^2} = -\frac{\gamma^*}{\sigma^2} < 0, \quad (42)$$

$$\frac{\partial S_B}{\partial \gamma^*} = \frac{(1-\theta)\phi D(t)}{r-r^*}\frac{1}{(1+\gamma^*)^2}, \quad (43)$$

and

$$\frac{\partial \left((S/S_B)^{-\gamma^*}\right)}{\partial \gamma^*} = \left(\frac{S}{S_B}\right)^{-\gamma^*}\frac{\partial}{\partial \gamma^*}\left(-\gamma^* \ln\left(\frac{S(r-r^*)(1+\gamma^*)}{(1-\theta)\phi D(t)\gamma^*}\right)\right)$$

$$= -\left(\frac{S}{S_B}\right)^{-\gamma^*}\left(\ln\left(\frac{S(r-r^*)(1+\gamma^*)}{(1-\theta)\phi D(t)\gamma^*}\right) + \gamma^*\left(\frac{1}{1+\gamma^*} - \frac{1}{\gamma^*}\right)\right) \quad (44)$$

$$= -\left(\frac{S}{S_B}\right)^{-\gamma^*}\left(\ln\left(\frac{S}{S_B}\right) - \frac{1}{1+\gamma^*}\right),$$

we have

$$\frac{\partial E(S)}{\partial \gamma^*} = -\frac{\partial S_B}{\partial \gamma^*}\left(\frac{S}{S_B}\right)^{-\gamma^*} + \left(\frac{(1-\theta)\phi D(t)}{r-r^*} - S_B\right)\cdot\frac{\partial}{\partial \gamma^*}\left(\left(\frac{S}{S_B}\right)^{-\gamma^*}\right)$$

$$= -\frac{(1-\theta)\phi D(t)}{r-r^*}\frac{1}{(1+\gamma^*)^2}\left(\frac{S}{S_B}\right)^{-\gamma^*}$$

$$-\left(\frac{(1-\theta)\phi D(t)}{r-r^*} - S_B\right)\left(\frac{S}{S_B}\right)^{-\gamma^*}\cdot\left(\ln\left(\frac{S}{S_B}\right) - \frac{1}{1+\gamma^*}\right) \quad (45)$$

$$= -\frac{(1-\theta)\phi D(t)}{r-r^*}\left(\frac{S}{S_B}\right)^{-\gamma^*}\frac{1}{1+\gamma^*}\ln\left(\frac{S}{S_B}\right).$$

Expression (45) is negative as long as $S > S_B$, that is, as long as bankruptcy hasn't been declared. Now, from (42), (45) implies:

$$\frac{\partial E(S)}{\partial \sigma^2} = \frac{\partial E(S)}{\partial \gamma^*}\frac{d\gamma^*}{d\sigma^2} > 0. \quad (46)$$

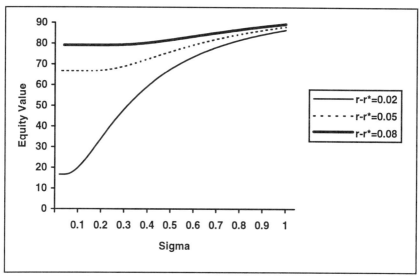

Figure 3.4: Plot of the value of equity against risk for different values of $r - r*$ and the following parameter values: $\theta = 1/3$, $\phi = 0.05$, $\alpha = 0.2$, $D(t) = 50$ and $S = 100$. As asset risk σ is increased, equity value rises, thus leading to a risk-shifting problem.

In words, the borrower has an incentive to increase project risk (Figure 3.4). This fact has an important implication for the optimal behavior of the lender: instead of monitoring asset *value*, he should focus on monitoring asset *risk,* since the latter is, in fact, the relevant variable for the bankruptcy decision of the borrower. Section 3.7 below demonstrates that, if project risk is given and known to the lender, then he can construct an incentive contract leading the agent to choose any bankruptcy strategy.

3.5.3 Measuring the Agency Cost of Debt: II

To measure the agency cost of debt due to risk-shifting, one must compare firm value in the social optimum with that achieved when equity holders are allowed to shift risk. From (25), it is not difficult to see that W will be maximal when $\gamma * \uparrow \infty$, i.e. when $\sigma \downarrow 0$. The intuition is the following: when asset risk vanishes, the value of the firm equals asset value plus the value of the tax shields. Because bankruptcy never occurs when $\sigma \downarrow 0$, the value of bankruptcy costs is zero.

Formally, the agency cost of debt due to risk shifting, C, is given by

$$C = \lim_{\gamma*\uparrow\infty} W(S) - \lim_{\gamma*\downarrow 0} W(S) . \tag{47}$$

Using

$$\lim_{\gamma*\uparrow\infty} W(S) = S + \frac{\theta\phi D(t)}{r-r*} \tag{48}$$

and

$$\lim_{\gamma*\downarrow 0} W(S) = S \tag{49}$$

yields the following value for the agency cost of debt, C:

$$C = \lim_{\gamma*\uparrow\infty} W(S) - \lim_{\gamma*\downarrow 0} W(S) = \frac{\theta\phi D(t)}{r-r*}. \tag{50}$$

Expression (50) implies that the value of the (safe) tax shields is lost if the borrower is allowed to engage in risk-shifting.[6]

3.5.4 The Incentive Effects of Loan Covenants

It may be interesting to analyze the incentive effects of loan covenants. Suppose that lender and borrower can agree on the following covenant: if asset value falls below a contractually pre-specified value $\overline{S}_B$, the firm will be liquidated. How does this covenant change the risk incentives faced by the borrower? Computing the partial derivative of equity value (26) with respect to σ^2, with the bankruptcy-triggering asset value $\overline{S}_B$ held constant, i.e. $\partial \overline{S}_B / \partial \sigma^2 = 0$ yields

$$\frac{\partial E(S)}{\partial \sigma^2} = \left(\frac{(1-\theta)\phi D(t)}{r-r*} - \overline{S}_B\right) \cdot \frac{\partial}{\partial\sigma^2}\left(\frac{S}{\overline{S}_B}\right)^{-\gamma*}$$

$$= \left(\frac{(1-\theta)\phi D(t)}{r-r*} - \overline{S}_B\right)\left(\frac{S}{\overline{S}_B}\right)^{-\gamma*} \ln\left(\frac{S}{\overline{S}_B}\right) \cdot \frac{\gamma*}{\sigma^2}. \tag{51}$$

The sign of expression (51) depends on that of

$$\left(\frac{(1-\theta)\phi D(t)}{r-r*} - \overline{S}_B\right). \tag{52}$$

If the bankruptcy trigger is low, so that (52) is positive, then the firm has a positive incentive to increase risk. If, however, (52) is negative, then equity value increases when risk is reduced. Thus, *a high enough safety covenant allows the lender to deter the borrower from taking undue risk.*

---

[6] Alternatively, if one were to assume that the range of asset risks available to equity holders is an interval $[\underline{\sigma}, \overline{\sigma}]$, the socially optimal project risk would be $\underline{\sigma}$, and equity holders would choose $\overline{\sigma}$. Hence, the agency cost of debt could be obtained by evaluating the expression $C = W(S; \sigma = \underline{\sigma}) - W(S; \sigma = \overline{\sigma})$.

More specifically, *when* $\bar{S}_B > (1-\theta)\phi D(t) / (r-r^*)$, *the borrower tries to reduce risk as much as possible to avoid bankruptcy.*

Note that any loan covenant $\bar{S}_B$ which is to be effective in deterring the borrower from shifting risk must lie *above* the endogenous bankruptcy-triggering asset value (28). Formally, for any asset risk σ, we have

$$\bar{S}_B > \frac{(1-\theta)\phi D(t)}{r-r^*} > \frac{(1-\theta)\phi D(t)}{r-r^*}\frac{\gamma^*}{1+\gamma^*} = S_B, \tag{53}$$

which means that a loan covenant that protects creditors against risk-shifting *also* leads shareholders to declare bankruptcy at a higher asset value. Therefore, we can say that loan covenants protect creditors in *two* ways. First, by triggering bankruptcy at a high enough asset value, they reduce losses incurred by creditors in the event of bankruptcy. Second, and more importantly, by rendering the bankruptcy-triggering asset value independent of firm asset risk, they mitigate or suppress the equity holders' risk-shifting incentives, thus avoiding the bankruptcy region ever being reached.[7]

3.6 The Financing Decision

3.6.1 Optimal Capital Structure

Using the results on endogenous bankruptcy, the equity holders' capital structure choice can now be analyzed. Again, assume that asset substitution is not possible, so that σ is known to the lender. Alternatively, we could assume that risk-shifting is, to a certain extent, possible but that the borrower cannot increase σ without bound. In this case, the lender would simply anticipate the borrower's subsequent risk-shifting behavior and use the *highest* asset risk achievable by the borrower in his calculations.

At the time the financing decision is made, the equity holders want to maximize the value of equity net of their initial investment I. The latter is equal to the current value of the project S minus outside financing received from the debt holders, which, in a competitive market, will equal $F(S)$. Hence, $I = S - F(S)$. Therefore, the equity holders maximize *net equity value*

$$E(S) - I = E(S) - (S - F(S)) = E(S) + F(S) - S. \tag{54}$$

---

[7] Note the similarity of this result with that obtained for collateral in Section 2.3.2.

This confirms the general result of agency theory that agency costs are eventually borne by the agent, here the equity holders.[8] Now, $E(S) + F(S) = W(S)$, so the equity holders actually maximize

$$W(S) - S = \frac{\theta \phi D(t)}{r - r^*} \left(1 - \left(\frac{S}{S_B} \right)^{-\gamma^*} \right) - \alpha S_B^{1+\gamma^*} S^{-\gamma^*} \qquad (55)$$

with respect to the face value of debt $D(t)$. Substituting for S_B (as given by (28)) in (55) yields

$$W(S) - S = \frac{\theta \phi D(t)}{r - r^*} \left(1 - \left(\frac{(1-\theta)\phi\gamma^* D(t)}{S(r - r^*)(1+\gamma^*)} \right)^{\gamma^*} \right)$$

$$- \alpha \left(\frac{(1-\theta)\phi\gamma^* D(t)}{(r - r^*)(1+\gamma^*)} \right)^{1+\gamma^*} S^{-\gamma^*} \qquad (56)$$

$$= \frac{\theta \phi D(t)}{r - r^*} - \left(D(t) \right)^{1+\gamma^*} S^{-\gamma^*} \cdot$$

$$\left(\frac{(1-\theta)\phi\gamma^*}{(r - r^*)(1+\gamma^*)} \right)^{\gamma^*} \frac{\theta + \gamma^*(\theta + \alpha(1-\theta))}{(r - r^*)(1+\gamma^*)} .$$

Net equity value (56) can be thought of as the profit accruing to equity holders when they organize the firm. The first-order condition for a maximum in (56) is[9]

$$\frac{\partial(W(S) - S)}{\partial D(t)} = \frac{\phi}{r - r^*} \left(\theta - \left(\frac{D(t)}{S} \frac{(1-\theta)\phi\gamma^*}{(r - r^*)(1+\gamma^*)} \right)^{\gamma^*} \right.$$

$$\left. \left(\theta + \gamma^*(\theta + \alpha(1-\theta)) \right) \right) = 0. \qquad (57)$$

Solving (57) for $D(t)$ yields the following expression for the optimal face value of debt, $\overline{D}(t)$:

---

[8] See Jensen and Meckling (1976), p. 91.
[9] It is a maximum, since

$$\frac{\partial^2(W(S) - S)}{\partial(D(t))^2} = -\frac{\gamma^*}{D(t)} \frac{\phi}{r - r^*} \left(\frac{D(t)}{S} \frac{(1-\theta)\phi\gamma^*}{(r - r^*)(1+\gamma^*)} \right)^{\gamma^*} .$$

$$\left(\theta + \gamma^*(\theta + \alpha(1-\theta)) \right) < 0.$$

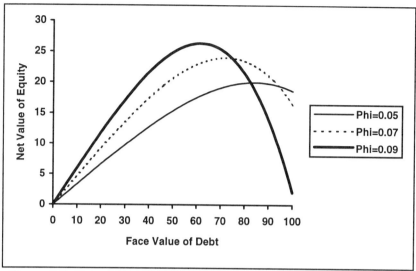

Figure 3.5: *Net equity value* $W(S) - S$ *as a function of the face value of debt* $D(t)$ *for different values of* ϕ *and the following parameter values:* $\theta = 1/3$, $\alpha = 0.2$, $S = 100$, $r - r^* = 0.05$ *and* $\sigma = 0.2$.

$$\overline{D}(t) = S\frac{(r - r^*)(1+\gamma^*)}{(1-\theta)\phi\gamma^*}\left(\frac{\theta}{\theta + \gamma^*(\theta + \alpha(1-\theta))}\right)^{1/\gamma^*}$$

$$= S\frac{r - r^* + \sigma^2/2}{(1-\theta)\phi}\left(\frac{\theta}{\theta + \gamma^*(\theta + \alpha(1-\theta))}\right)^{1/\gamma^*}. \quad (58)$$

Figures 3.5 and 3.6 show plots of net equity value $W(S) - S$ for different face values of debt for a set of parameters. In both figures, net equity value $W(S) - S$ reaches an interior maximum, illustrating the generic existence of an optimal capital structure balancing tax benefits and bankruptcy costs. From equation (58), it is easy to see that the optimal face value of debt is decreasing in the interest rate ϕ:

$$\frac{\partial \overline{D}(t)}{\partial \phi} = -S\frac{r - r^* + \sigma^2/2}{(1-\theta)\phi^2}\left(\frac{\theta}{\theta + \gamma^*(\theta + \alpha(1-\theta))}\right)^{1/\gamma^*} < 0. \quad (59)$$

This result is depicted in Figure 3.7, which plots the optimal face value of debt as a function of the interest rate ϕ. An additional insight into the properties of optimal capital structure can be gained from (58). Multiplying this expression by ϕ, we can see that the optimal instantaneous *coupon* payment

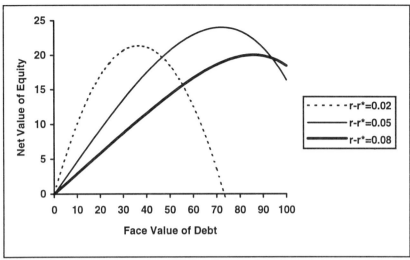

Figure 3.6: Net equity value $W(S) - S$ *as a function of the face value of debt* $D(t)$ *for different values of* $r - r*$ *and the following parameter values:* $\theta = 1/3$, $\phi = 0.07$, $\alpha = 0.2$, $S = 100$ *and* $\sigma = 0.2$.

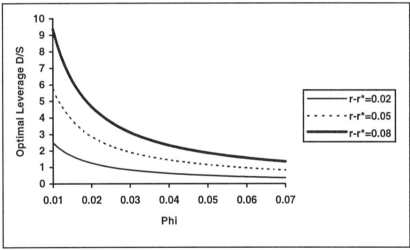

Figure 3.7: Optimal leverage $\overline{D}/S$ *for different values of* $r - r*$ *and the following parameter values:* $\theta = 1/3$, $\alpha = 0.2$ *and* $\sigma = 0.2$. *As the interest rate* ϕ *is increased, optimal leverage falls.*

$$\phi\overline{D}(t) = S\frac{r - r* + \sigma^2/2}{1 - \theta}\left(\frac{\theta}{\theta + \gamma*(\theta + \alpha(1 - \theta))}\right)^{1/\gamma*} \qquad (60)$$

is independent of ϕ.

The sign of $\partial \overline{D}(t) / \partial(r - r^*)$ is difficult to ascertain analytically. A numerical computation for the base case shows that it is quite likely to be negative, as is illustrated in Figure 3.8.

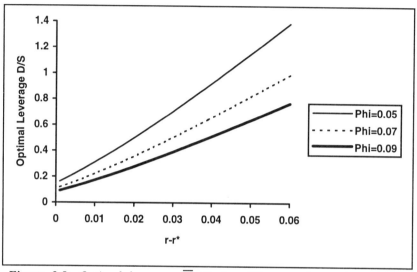

Figure 3.8: *Optimal leverage* $\overline{D} / S$ *for different values of* ϕ *and the following parameter values:* $\theta = 1 / 3$, $\alpha = 0.2$ *and* $\sigma = 0.2$. *As the interest rate* r^* *is increased, optimal leverage falls.*

3.6.2 Interest Payments vs. Increase in the Face Value of Debt

Using the optimal capital structure result above, the question of how debt service is going to be split between r^* and ϕ can now be addressed. To do this, substitute the result (58) in the equation for $W(S) - S$ to get

$$W(S) - S = \frac{S}{\vartheta}\left(1 + \gamma *\left(1 + \alpha \vartheta\right)\right)^{-1/\gamma*}, \quad \vartheta \equiv \frac{1 - \theta}{\theta}. \tag{61}$$

Inspection of equation (61) shows that $W(S) - S$ does not depend on ϕ. Hence, once account is taken of the borrower's capital structure choice, he is *indifferent* as to which interest rate he effectively has to pay to the lender. Such is not the case, however, for the interest rate differential $r - r^*$. To see this, recall that

$$\frac{d\gamma *}{d(r - r^*)} = \frac{2}{\sigma^2}. \tag{62}$$

Then,

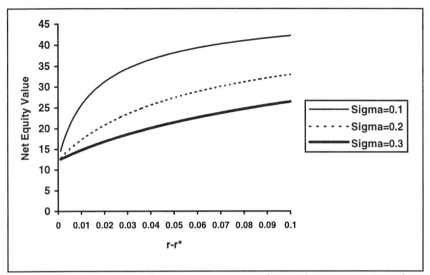

Figure 3.9: *Influence of the interest rate spread* $r - r*$ *on net equity value* $W(S) - S$ *for the following parameter values:* $\theta = 1/3$, $\alpha = 0.2$ *and* $\phi = 0.07$. *As* $r - r*$ *is increased, net equity value rises, implying that an increase in r* reduces net equity value.*

$$\frac{\partial(W(S)-S)}{\partial(r-r*)} = \frac{\partial(W(S)-S)}{\partial\gamma*} \cdot \frac{d\gamma*}{d(r-r*)}$$

$$= \frac{S(1+\gamma*(1+\alpha\vartheta))^{-1/\gamma*}}{\vartheta(r-r*)}\left(\frac{\ln(1+\gamma*(1+\alpha\vartheta))}{\gamma*} - \frac{1+\alpha\vartheta}{1+\gamma*(1+\alpha\vartheta)}\right). \tag{63}$$

One can show that (63) is positive, which means that net equity value rises with the interest rate spread $r - r*$, as Figure 3.9 illustrates.[10] This in turn

---

[10] For $\gamma* = 0$, the value of the expression

$$\left(\ln(1+\gamma*(1+\alpha\vartheta)) - \frac{\gamma*(1+\alpha\vartheta)}{1+\gamma*(1+\alpha\vartheta)}\right),$$

which determines the sign of (63), is zero. Now, as $\gamma*$ increases, this expression becomes positive, since

$$\frac{\partial}{\partial\gamma*}\left(\ln(1+\gamma*(1+\alpha\vartheta)) - \frac{\gamma*(1+\alpha\vartheta)}{1+\gamma*(1+\alpha\vartheta)}\right)$$

$$= \frac{1+\alpha\vartheta}{1+\gamma*(1+\alpha\vartheta)}\left(1 - \frac{1}{1+\gamma*(1+\alpha\vartheta)}\right) > 0$$

for $\gamma* > 0$. Thus, expression (63) is positive as well.

implies that net equity value $W(S) - S$ is *decreasing* in the interest rate
$r*$. We therefore have the following result: the borrower is indifferent as
to which ϕ is to be chosen, but strongly prefers a low $r*$. Hence, we should
expect debt contracts to call for effective payments, instead of adding the
interest to the principal amount of the debt.[11]

3.6.3 Equilibrium on the Credit Market

The results presented above suggest an interesting implication for
equilibrium on the credit market. To understand this point, assume that,
although the market is competitive, there is a *given* amount of funds
available for lending purposes.

There are basically two ways in which equilibrium on quantities can be
attained: either through an increase in ϕ or through an increase in $r*$.
Although any increase in ϕ or $r*$ leads, through the capital structure
choice, to a reduction in the optimal *face* value of debt $\overline{D}$, increasing the
interest rate ϕ has no influence on the amount of external financing F
actually collected by shareholders. To see this, substitute (58) in (28) and
(15) to get

$$F(S) = \frac{S}{1-\theta} \frac{(1+\alpha\vartheta)(1+\gamma^*)+(1-\alpha)(1-\theta)}{(1+\gamma^*(1+\alpha\vartheta))^{1+1/\gamma^*}}, \qquad (64)$$

which does not depend on ϕ. Therefore, clearing on the credit market must
be achieved through the interest rate $r*$.

3.6.4 Capital Structure and the Expected Life of Companies

The results obtained here can provide an interesting insight into the
expected life of companies as a function of capital structure and interest
payments. To analyze this problem, recall that, if capital structure is
optimally chosen at initial time 0, then we have by the previous analysis

$$\overline{D}_0 = S_0 \frac{r - r^* + \sigma^2/2}{(1-\theta)\phi} \left(\frac{\theta}{\theta + \gamma^*(\theta + \alpha(1-\theta))} \right)^{1/\gamma^*}. \qquad (65)$$

Now, bankruptcy occurs when

$$S_B = \frac{(1-\theta)\phi D(t)}{r - r^* + \sigma^2/2}. \qquad (28)$$

---

[11] Note that this is true at the time the contract is signed. Afterwards, once the
amount of debt has been chosen, the borrower might display different preferences.

Substituting the optimal capital structure at initial time (65) into (28) yields

$$S_B(t) = \frac{(1-\theta)\phi}{r - r^* + \sigma^2/2} \overline{D}_0 e^{r^* t}$$

$$= S_0 e^{r^* t} \left(\frac{1}{1 + \gamma^*(1 + \alpha\vartheta)} \right)^{1/\gamma^*}. \tag{66}$$

From the analysis in Ingersoll (1987, p. 353), the mean time to absorbtion at the origin for a random variable following the process $dx = \mu dt + \sigma d\tilde{z}$ with initial value x_0 is given by

$$\bar{\tau} = \frac{x_0}{\mu}. \tag{67}$$

Now, $S/S_B(t)$ is a lognormally distributed random variable with initial value

$$\frac{S_0}{S_B(0)} = \frac{S_0}{S_0 \left(\dfrac{1}{1 + \gamma^*(1 + \alpha\vartheta)} \right)^{1/\gamma^*}} = \left(1 + \gamma^*(1 + \alpha\vartheta)\right)^{1/\gamma^*} \tag{68}$$

and drift[12]

$$\mu - r^* - \frac{\sigma^2}{2}. \tag{69}$$

Hence, taking logarithms, we obtain

$$x_0 = \ln \frac{S_0}{S_B(0)} = \frac{1}{\gamma^*} \ln\left(1 + \gamma^*(1 + \alpha\vartheta)\right), \tag{70}$$

and the mean time to absorbtion (i.e. bankruptcy) equals

$$\bar{\tau} = \frac{x_0}{\mu - r^* - \sigma^2/2} = \frac{\ln\left(1 + \gamma^*(1 + \alpha\vartheta)\right)}{\gamma^*\left(\mu - r^* - \sigma^2/2\right)}. \tag{71}$$

Inspection of equation (71) shows that the mean time to bankruptcy does *not* depend on the parameter ϕ. This means that the borrower, if allowed to choose his capital structure freely, will make his choice so as to keep the mean time to bankruptcy *constant*. Again, this is not the case for the parameter r^*.

---

[12] Let a random variable S follow a geometric Brownian motion. Then, $\ln S$ has drift $\mu - \sigma^2/2$ and variance σ^2. See Hull (1993), p. 209.

3.7 An Incentive Contract

By choosing the appropriate value for the interest payment parameter ϕ, the lender can actually lead the borrower to select any bankruptcy-triggering value S_B. In this section, we develop an *incentive contract* in which the lender sets ϕ such that he suffers no (nominal) losses upon bankruptcy. In other words, ϕ is set such that the borrower declares bankruptcy when the asset value reaches $D(t)/(1-\alpha)$, thus allowing the lender to recover the *face amount* of his claim, $D(t)$.

Such a policy might be interesting for the lender for a number of reasons, which have their roots in the fact that most of real-world institutions operate with nominal amounts. First, he might himself be an agent and fear being sued for taking too much risk (which, in the real world, occurs when loans are not repaid "in full" in nominal terms). Second, liquidating assets early may increase confidence in his solvability and avoid some problems typical for financial intermediaries, such as bank runs. Furthermore, this policy enables the lender to avoid having to monitor the borrower and to save on monitoring costs, since he can use the (early) information provided by default on interest payments to engage in project liquidation and recover the principal amount of his debt, $D(t)$. Finally, as the analysis on capital structure has demonstrated, varying ϕ merely leads the borrower to borrow less and has no influence on the value of the firm. In other words, the incentive contract has no social welfare costs.[13]

To lead the agent to default as soon as the asset value reaches $D(t)/(1-\alpha)$, the lender should set ϕ such that

$$S_B = \frac{(1-\theta)\phi D(t)}{r-r^*}\frac{\gamma^*}{1+\gamma^*} = \frac{D(t)}{1-\alpha}, \tag{72}$$

yielding an "optimal" ϕ:

$$\bar{\phi} = \frac{r-r^*}{(1-\alpha)(1-\theta)}\frac{1+\gamma^*}{\gamma^*} = \frac{(r-r^*)+\sigma^2/2}{(1-\alpha)(1-\theta)}. \tag{73}$$

It can be interesting to analyze the properties of this incentive contract. Its most striking feature is to result in a total debt service that lies *above* the risk-free rate of return.[14] To see this, consider the total instantaneous rate

---

[13] In effect, since the optimal instantaneous coupon payment is independent of ϕ, the contract has no effect on the mean time to bankruptcy either. Bankruptcy occurs early in the sense that it is declared when asset value reaches a value that lies above the face value of debt.

[14] By assumption, however, the loan is fairly priced.

of return, which is the sum of the effective interest payment rate ϕ and the rate of increase in the face value of debt $r*$:

$$\bar{\phi} + r* = \frac{(r - r*) + \sigma^2 / 2}{(1-\alpha)(1-\theta)} + r* = \frac{r - (\theta(1-\alpha) + \alpha)r* + \sigma^2 / 2}{(1-\alpha)(1-\theta)}. \quad (74)$$

Taking the partial derivative of (74) with respect to $r*$, we get

$$\frac{\partial(\bar{\phi} + r*)}{\partial r*} = -\frac{\theta(1-\alpha) + \alpha}{(1-\alpha)(1-\theta)} = 1 - \frac{1}{(1-\alpha)(1-\theta)} < 0. \quad (75)$$

Equation (75) means that total payments are reduced whenever the growth rate in $D(t)$, $r*$, is raised. Assume we would like total debt service to equal the risk-free rate of return r. Then, we must have

$$\bar{\phi} + r* = \frac{(r - r*) + \sigma^2 / 2}{(1-\alpha)(1-\theta)} + r* = r$$

$$\Leftrightarrow \quad (r - r*)\big(1 - (1-\alpha)(1-\theta)\big) = -\frac{\sigma^2}{2}, \quad (76)$$

that is

$$r* = r + \frac{\sigma^2}{2} \frac{1}{1 - (1-\alpha)(1-\theta)} > r. \quad (77)$$

But then,

$$\bar{\phi} = \frac{r - r* + \sigma^2 / 2}{(1-\alpha)(1-\theta)} = -\frac{\sigma^2}{2} \frac{1}{1 - (1-\alpha)(1-\theta)} < 0. \quad (78)$$

If $\bar{\phi} < 0$, however, the borrower never defaults and the incentive mechanism breaks down. We therefore must have $\bar{\phi} > 0$. This condition then implies

$$\bar{\phi} = \frac{(r - r*) + \sigma^2 / 2}{(1-\alpha)(1-\theta)} > 0 \quad \Leftrightarrow \quad r* < r - \frac{\sigma^2}{2}. \quad (79)$$

Therefore, we get

$$\bar{\phi} + r* = \frac{(r - r*) + \sigma^2 / 2}{(1-\alpha)(1-\theta)} + r* = \frac{r - r*\big(1 - (1-\alpha)(1-\theta)\big) + \sigma^2 / 2}{(1-\alpha)(1-\theta)}$$

$$> \frac{r - (r - \sigma^2 / 2)\big(1 - (1-\alpha)(1-\theta)\big) + \sigma^2 / 2}{(1-\alpha)(1-\theta)} \quad (80)$$

$$= \frac{(r - \sigma^2 / 2)(1-\alpha)(1-\theta)}{(1-\alpha)(1-\theta)} = r + \frac{\sigma^2}{2}.$$

In order to maintain the incentive mechanism, the lender must set $\bar{\phi} > 0$.
But then, he must ask for a total interest payment $\bar{\phi} + r^* > r + \sigma^2/2$. That
is, he asks for a total interest above the risk-free rate.[15]

3.8 Extensions of the Model

The analysis presented above can be extended to other payout settings. As
an example, consider the case in which total payouts are a fixed proportion
β of total asset value, to be shared between lender and borrower. Then, S
evolves according to

$$dS = (\mu - \beta)Sdt + \sigma Sd\tilde{z}. \tag{81}$$

In this setting, the lender receives $\phi D(t)$ and the borrower $\beta S_t - \phi D(t)$
per unit time.

3.8.1 The Value of the Firm and its Securities

The value of the debt satisfies the following partial differential equation:

$$\tfrac{1}{2}\sigma^2 S^2 F_{SS} + (r - \beta)SF_S + r^* D(t)F_D - rF + \phi D(t) = 0. \tag{82}$$

Making the same change in variables as in Section 3.3.1, i.e.
$V = S_t / D(t)$, defining $G(V) = F(S)/D(t)$, and using the same line of
reasoning that lead to equation (5), we get

$$\tfrac{1}{2}\sigma^2 V^2 G'' + (r - r^* - \beta)VG' - (r - r^*)G + \phi = 0, \tag{83}$$

which has general solution

$$G = \alpha_0 + \alpha_1 V^{\lambda_1} + \alpha_2 V^{\lambda_2}, \tag{84}$$

where

$$\lambda_1 = -\frac{b - \sqrt{b^2 + 2(r - r^*)\sigma^2}}{\sigma^2} > 0, \tag{85}$$

$$\lambda_2 = -\frac{b + \sqrt{b^2 + 2(r - r^*)\sigma^2}}{\sigma^2} < 0, \tag{86}$$

$$b \equiv r - r^* - \beta - \sigma^2/2. \tag{87}$$

Multiplying by $D(t)$, we get

---

[15] As an alternative way of seeing that $\phi + r^* = r$ yields a trigger value smaller
than the principal amount of the loan, just set $\phi = r - r^*$ in equation (28) to
obtain

$$S_B = (1 - \theta)D(t)\frac{\gamma^*}{1 + \gamma^*} < D(t).$$

$$F = \alpha_0 D(t) + \alpha_1 S^{\lambda_1} D(t)^{1-\lambda_1} + \alpha_2 S^{\lambda_2} D(t)^{1-\lambda_2}. \tag{88}$$

This value must satisfy the same boundary conditions as in the base case, i.e.

$$F(S_B) = (1-\alpha)S_B, \tag{9}$$

$$F(\infty) = \frac{\phi}{r - r^*} D(t). \tag{10}$$

From boundary condition (10), we get $\alpha_1 = 0$, and then

$$\alpha_0 = \frac{\phi}{r - r^*}, \tag{89}$$

$$\alpha_2 = \left(\frac{S_B}{D(t)}\right)^{-\lambda_2}\left((1-\alpha)\frac{S_B}{D(t)} - \frac{\phi}{r - r^*}\right). \tag{90}$$

Hence, the value of debt, F, equals:

$$F(S) = D(t)\left(\frac{\phi}{r - r^*} + \left((1-\alpha)\frac{S_B}{D(t)} - \frac{\phi}{r - r^*}\right)\left(\frac{S}{S_B}\right)^{\lambda_2}\right), \tag{91}$$

which is just equation (15) above with $-\gamma^*$ replaced by λ_2. Analogously, the value of bankruptcy costs is easily seen to be

$$K(S) = \alpha S_B^{1-\lambda_2} S^{\lambda_2}, \tag{92}$$

and the value of the tax shields

$$TB(S) = \frac{\theta\phi D(t)}{r - r^*}\left(1 - \left(\frac{S}{S_B}\right)^{\lambda_2}\right). \tag{93}$$

Then, the value of equity, E, is:

$$E(S) = S - \frac{(1-\theta)\phi D(t)}{r - r^*}\left(1 - \left(\frac{S}{S_B}\right)^{\lambda_2}\right) - S_B^{1-\lambda_2} S^{\lambda_2}. \tag{94}$$

3.8.2 The Bankruptcy Decision

From (94), the bankruptcy-triggering asset value S_B can be computed as:

$$S_B = -\frac{(1-\theta)\phi D(t)}{r - r^*}\frac{\lambda_2}{1 - \lambda_2}. \tag{95}$$

Therefore, the contract that yields net proceeds from bankruptcy equal to the principal amount of the loan is given by

$$\overline{\phi} = -\frac{r - r^*}{(1-\alpha)(1-\theta)}\frac{1 - \lambda_2}{\lambda_2}. \tag{96}$$

3.8.3 The Effect of the Payout Rate on Equity Value

An interesting question that arises in this new setting is that of the effect of an increase in the payout rate β on the value of equity. Using

$$\frac{\partial S_B}{\partial \lambda_2} = -\frac{(1-\theta)\phi D(t)}{r-r^*}\frac{1}{(1-\lambda_2)^2} < 0 \tag{97}$$

and

$$\frac{\partial (S/S_B)^{\lambda_2}}{\partial \lambda_2} = \frac{\partial}{\partial \lambda_2}\left(\frac{S}{S_B}\right)^{\lambda_2} = \frac{\partial}{\partial \lambda_2}\left(-\frac{S(r-r^*)(1-\lambda_2)}{(1-\theta)\phi D(t)\lambda_2}\right)^{\lambda_2}$$

$$= \left(\frac{S}{S_B}\right)^{\lambda_2}\cdot\left(\ln\left(\frac{S}{S_B}\right) - \frac{1}{1-\lambda_2}\right), \tag{98}$$

we get

$$\frac{\partial E(S)}{\partial \lambda_2} = \frac{(1-\theta)\phi D(t)}{r-r^*}\frac{\partial}{\partial \lambda_2}\left(\frac{S}{S_B}\right)^{\lambda_2}$$

$$-\left(\frac{\partial S_B}{\partial \lambda_2}\left(\frac{S}{S_B}\right)^{\lambda_2} + S_B\frac{\partial}{\partial \lambda_2}\left(\frac{S}{S_B}\right)^{\lambda_2}\right) \tag{99}$$

$$= \frac{(1-\theta)\phi D(t)}{r-r^*}\left(\frac{S}{S_B}\right)^{\lambda_2}\frac{1}{1-\lambda_2}\ln\left(\frac{S}{S_B}\right)$$

Now,

$$\frac{d\lambda_2}{d\beta} = \frac{1}{\sigma^2}\left(1 + \frac{r-r^*-\beta-\sigma^2/2}{\sqrt{(r-r^*-\beta-\sigma^2/2)^2 + 2(r-r^*)\sigma^2}}\right) > 0, \tag{100}$$

so the derivative of the value of equity with respect to the payout rate,

$$\frac{\partial E(S)}{\partial \beta} = \frac{\partial E(S)}{\partial \lambda_2}\frac{d\lambda_2}{d\beta}, \tag{101}$$

is positive as long as bankruptcy has not been declared. Hence, shareholders can raise the value of their equity claim by *maximizing* payouts. The reason is that by doing so, they take money away from the creditors. This effect is depicted in Figure 3.10. In the limit, shareholders might decide to pay out the whole asset value immediately, i.e. set $\beta \uparrow \infty$. One can show that in this case, equity value equals asset value S, i.e.

$$\lim_{\beta\uparrow\infty} E(S) = S. \tag{102}$$

To protect himself against excessive payouts, the lender will therefore have an incentive to enforce a loan covenant limiting payouts.

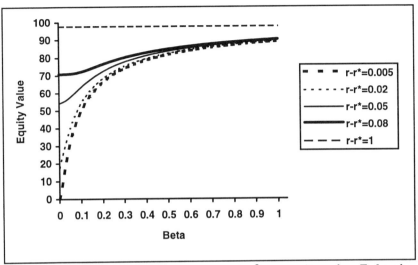

Figure 3.10: Influence of the payout rate β on equity value E for the following parameter values: $\theta = 1/3$, $\phi = 0.05$, $\alpha = 0.2$, $D(t) = 70$, $S = 100$ and $\sigma = 0.2$. As the payout rate is increased, the value of equity rises.

3.8.4 Effect of a Loan Covenant on the Optimal Payout Rate

The shareholders' incentive to increase payouts is very similar to the risk-shifting problem discussed in Section 3.5.2. Therefore, in addition to a *direct* covenant specifying a maximum payout rate, debt holders can use an *indirect* covenant of the form presented in Section 3.5.4. to mitigate the shareholders' incentive to increase payouts. To see this, suppose that a loan covenant setting a pre-specified bankruptcy-triggering asset value $\overline{S}_B$ is agreed upon. With $\overline{S}_B$ fixed,

$$\frac{\partial E(S)}{\partial \lambda_2} = \left(\frac{(1-\theta)\phi D(t)}{r - r*} - \overline{S}_B \right) \left(\frac{S}{\overline{S}_B} \right)^{\lambda_2} \ln\left(\frac{S}{\overline{S}_B} \right). \tag{103}$$

Using the fact that $\partial E(S) / \partial \beta = (\partial E(S) / \partial \lambda_2) \cdot (d\lambda_2 / d\beta)$ and $d\lambda_2 / d\beta > 0$, the shareholders will have an incentive to increase payouts whenever

$$\left(\frac{(1-\theta)\phi D(t)}{r-r*} - \bar{S}_B \right) > 0 . \tag{104}$$

If $\bar{S}_B > (1-\theta)\phi D(t) / (r-r*)$, however, the shareholders will seek to minimize payouts. Again, note the similarity of this result with that on the effect of a loan covenant on the equity holders' risk-shifting incentives.

We are now able to state the following general result: if the asset value is observable, then lender and borrower can solve *both* the risk-shifting problem and the problem of excessive payouts by agreeing on a covenant specifying that liquidation will be triggered as soon as the asset value reaches a pre-specified value $\bar{S}_B$ such that $\bar{S}_B > (1-\theta)\phi D(t) / (r-r*)$. Risk-shifting and excessive payouts are to be expected, however, if the asset value is not observable or such a covenant is not enforceable. In this case, the lender should concentrate on monitoring asset risk and payouts, since these are, in fact, the relevant factors influencing the shareholders' bankruptcy decision and the value of debt. In this respect, monitoring the asset value on the one hand and monitoring asset risk and payouts on the other can be considered as *substitutes*.

3.9 Conclusion

In this chapter, we used a simple model to analyze the incentive effects of loan contracts. First, the equity holders' *bankruptcy* decision was analyzed. More specifically, it was demonstrated that, with endogenous bankruptcy,

- lender and equity holders will, in general, not agree as to when bankruptcy is to be declared.
- capital structure and interest rates have an influence on the borrower's bankruptcy decision. The asset value at which the borrower defaults on interest payments rises with the face value of debt and the interest rate effectively paid on debt and falls with asset risk.
- the borrower's bankruptcy decision is socially suboptimal in the sense that it does not maximize overall firm value. In other words, endogenous bankruptcy creates a principal-agent problem. The resulting agency cost of debt can be measured. It is increasing in the face value of debt, the interest rate effectively paid on debt and bankruptcy costs.
- some common characteristics of loan contracts, such as a positive effective interest rate and the provision that the loan can be called only when the borrower defaults on the interest payment, can be derived endogenously as a consequence of the conflict of interest that exists between lenders and equity holders.

Turning to the firm's *investment* decision, it was then shown that Myers' underinvestment problem exists: because of the existence of debt, firms forgo profitable investment opportunities. The reason is that when shareholders recapitalize the firm, a part of the resulting increase in overall firm value accrues to debt holders, making the additional investment unprofitable.

The existence of debt gives shareholders an incentive to increase project risk in order to raise the value of equity. The agency cost resulting from this second distortion of the firm's investment decision can be measured. The risk-shifting problem can be overcome through the use of a bond covenant specifying that liquidation is to occur if asset value falls below a contractually pre-specified level. Enforcement of this covenant requires that the lender monitors asset value carefully.

Extending the model to the case of endogenous payouts, it was shown that the borrower's incentive to increase the payout rate is analytically very similar to the risk-shifting problem. It can be overcome through the use of a bond covenant specifying that liquidation is to occur if asset value falls below a contractually pre-specified level. As in the case of risk-shifting, enforcement of this covenant requires that the lender monitors asset value carefully.

Alternatively, the lender can limit payouts contractually and monitor asset risk carefully to avoid having to monitor the asset value. In this respect, monitoring the asset value on the one hand and monitoring asset risk and payouts on the other can be considered as substitutes. The reason is that when project risk and the payout rate are given, the equity holders' bankruptcy strategy is fully determined by contractual provisions. Moreover, by setting an appropriate interest rate, an incentive contract can be constructed that leads the borrower to declare bankruptcy once asset value reaches a certain share of the face value of debt.

In analyzing the firm's *capital structure* choice, it was shown that a rising interest rate implies a lower optimal nominal leverage but has no influence on the amount of outside financing actually collected by shareholders by selling debt. At the time of the financing, the optimal instantaneous coupon payment, overall firm value, and mean time to bankruptcy are all independent of the interest rate.

4. Junior Debt

4.1 Introduction

While many papers in the literature have provided a rationale for the existence of debt, such as tax benefits or signaling effects, only a few have made the distinction between senior and junior debt and analyzed the consequences of differing priority of claims for firm behavior. Perotti and Spier (1993) and Hart and Moore (1995) are two recent exceptions. Both of these papers explore the effects of the existence of both senior and junior debt on the firm's investment decision. Perotti and Spier (1993) show that value may be extracted from senior claims through the issue of junior debt. The reason is that by retiring equity through a junior debt issue, the shareholders credibly threaten not to undertake valuable new investment unless senior debt holders concede to a reduction of their claims. Hart and Moore (1995) show that a mix of short-term and senior long-term debt might be necessary to deter management from undertaking unprofitable investment. The basic intuition, extending that of Jensen (1986), is that short-term debt forces management to disgorge free cash flows.

This chapter uses the game theory analysis of options to analyze the incentive effects of junior debt with the firm's investment strategy held *constant*. Drawing on the model of Chapter 3, we price senior and junior debt, show that the existence of senior debt distorts the equity holders' choice to issue junior debt and that a junior debt issue has a negative influence on the value of senior debt, thus invalidating the conventional wisdom that seniority fully protects debt holders against adverse wealth changes resulting from the issue of new debt. This extends the results of Perotti and Spier (1993) to the case of a constant investment strategy.

4.2 The Model

The model used in this chapter draws heavily on that of Chapter 3. As before, the value of the borrower's assets, S, is assumed to follow the geometric Brownian motion

$$dS = \mu S dt + \sigma S d\tilde{z} . \tag{1}$$

Asset substitution is not possible, so that the parameters μ and σ are known to all parties. As was the case in the previous chapter, assume that asset sales are prohibited. Hence, any net cash outflows associated with interest payments must be financed by selling additional equity.

Assume further that the borrower, to finance his project, borrows from a first lender, 1, with whom he reaches the following agreement: in exchange of a loan of F_1, the borrower is to pay an instantaneous interest of $\phi_1 D_1 dt$ to the lender, where D_1 and ϕ_1 denote the face value of debt and the interest rate, respectively.

Finally, assume that the borrower is free to borrow additional amounts from other lenders, but that any other claim issued will be junior to the first claim. For simplicity, we will assume that the borrower only borrows once more; the analysis could be extended to a higher number of claims, at the cost of greater complexity. Let this second claim be denoted by F_2 and consist of a promise by the borrower to pay to lender 2 an instantaneous interest of $\phi_2 D_2 dt$, the loan having a face amount of D_2.

If (and only if) the borrower defaults on any of the promised payments, the firm will be considered bankrupt vis-à-vis all the creditors and liquidated. If bankruptcy occurs, a fraction $0 \le \alpha < 1$ of value is lost, leaving creditors with $(1-\alpha)S_B$, where S_B denotes the asset value at which bankruptcy occurs. The structure of this game is depicted in Figure 4.1.

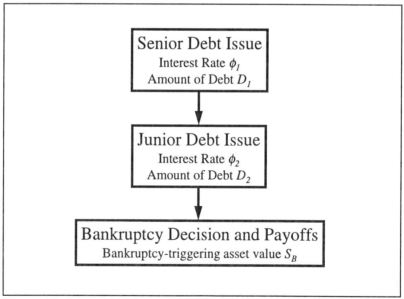

Figure 4.1: *Structure of the game. In the first phase, senior debt is issued. In a later period, equity holders might decide to issue junior debt. Finally, if the equity holders default on interest payments, bankruptcy is declared and the firm is liquidated.*

The structure of the chapter is as follows: Section 4.3 uses option pricing to determine the value of the firm and its securities. Section 4.4 analyzes the equity holders' bankruptcy decision. Section 4.5 explores the firm's decision to issue junior debt and shows that it is distorted by the existence of senior debt. Section 4.6 describes the influence of a junior debt issue on the value of senior debt and illustrates that seniority need not protect debt holders against adverse wealth changes resulting from the issue of new debt. Section 4.7 concludes the chapter.

4.3 The Value of the Firm and its Securities

Once the game has been specified, the next step in the method is to value the players' payoffs using option pricing theory, treating all the players' decision variables as parameters. This is done in this section.

4.3.1 The Value of Senior Debt

The value of the senior lender's claim, F_1, satisfies the following differential equation:

$$\tfrac{1}{2}\sigma^2 S^2 F_1'' + rSF_1' - rF_1 + \phi_1 D_1 = 0, \tag{2}$$

which has general solution

$$F_1 = \alpha_0 + \alpha_1 S + \alpha_2 S^{-\gamma}, \quad \gamma \equiv 2r / \sigma^2. \tag{3}$$

This value must satisfy the following boundary conditions:

$$F_1(S_B) = Min[(1-\alpha)S_B; D_1], \tag{4}$$

$$F_1(\infty) = \frac{\phi_1 D_1}{r}. \tag{5}$$

Equation (4) stems from the properties of the bankruptcy process; equation (5) states that bankruptcy becomes irrelevant as S becomes very large. From (5), α_1 must be zero. Hence, we can write

$$F_1 = \alpha_0 + \alpha_2 S^{-\gamma}. \tag{6}$$

Then, from (5), we have

$$\alpha_0 = \frac{\phi_1 D_1}{r}. \tag{7}$$

The problem now is that the asset value that triggers bankruptcy, S_B, is still unknown, and we therefore do not know which of the values in (4) to use. Considering the case where $(1-\alpha)S_B < D_1$ first and using (4), we obtain

$$F_1(S_B) = (1-\alpha)S_B = \frac{\phi_1 D_1}{r} + \alpha_2 S_B^{-\gamma} \tag{8}$$

and

$$\alpha_2 = \left((1-\alpha)S_B - \frac{\phi_1 D_1}{r} \right) S_B^\gamma . \tag{9}$$

Therefore, the value of senior debt, F_1, equals:

$$F_1(S) = \frac{\phi_1 D_1}{r} + \left((1-\alpha)S_B - \frac{\phi_1 D_1}{r} \right) \left(\frac{S}{S_B} \right)^{-\gamma} . \tag{10}$$

In the second case, where $(1-\alpha)S_B < D_1$, boundary condition (4) yields

$$F_1(S_B) = D_1 = \frac{\phi_1 D_1}{r} + \alpha_2 S_B^{-\gamma} \tag{11}$$

and

$$\alpha_2 = D_1 \left(1 - \frac{\phi_1}{r} \right) S_B^\gamma . \tag{12}$$

Therefore, the value of senior debt, F_1, is given by:

$$F_1(S) = \frac{\phi_1 D_1}{r} + D_1 \left(1 - \frac{\phi_1}{r} \right) \left(\frac{S}{S_B} \right)^{-\gamma} . \tag{13}$$

These results can be summarized as follows:

$$F_1(S) = \begin{cases} \dfrac{\phi_1 D_1}{r} + \left((1-\alpha)S_B - \dfrac{\phi_1 D_1}{r} \right) \left(\dfrac{S}{S_B} \right)^{-\gamma} & \text{if } (1-\alpha)S_B < D_1 \\[4mm] \dfrac{\phi_1 D_1}{r} + D_1 \left(1 - \dfrac{\phi_1}{r} \right) \left(\dfrac{S}{S_B} \right)^{-\gamma} & \text{if } (1-\alpha)S_B > D_1. \end{cases} \tag{14}$$

As in Chapter 3, (15) can be interpreted in terms of the risk-neutral probability of bankruptcy, $(S/S_B)^{-\gamma}$. In order to do so, rewrite senior debt value as $F_1(S) = (\phi_1 D_1 / r)(1 - (S/S_B)^{-\gamma}) + (1-\alpha)S_B(S/S_B)^{-\gamma}$ for $(1-\alpha)S_B < D_1$ and as $F_1(S) = (\phi_1 D_1 / r)(1 - (S/S_B)^{-\gamma}) + D_1(S/S_B)^{-\gamma}$ for $(1-\alpha)S_B > D_1$. Then, the value of senior debt equals the value of the risk-free debt, $\phi_1 D_1 / r$, times the risk-neutral probability that bankruptcy does not occur, $1 - (S/S_B)^{-\gamma}$, plus the payoff to senior creditors in the event of bankruptcy, $(1-\alpha)S_B$ or D_1 (depending on whether $(1-\alpha)S_B < D_1$ or $(1-\alpha)S_B > D_1$), times the risk-neutral probability of bankruptcy, $(S/S_B)^{-\gamma}$.

4.3.2   The Value of Junior Debt

A similar analysis can be conducted to determine the value of junior debt, F_2. It must also satisfy equation (2), that is, F_2 must be of the form

$$F_2 = \alpha_0 + \alpha_1 S + \alpha_2 S^{-\gamma}, \tag{15}$$

subject to the boundary conditions

$$F_2(S_B) = Max\left[0; Min\left[(1-\alpha)S_B - D_1; D_2\right]\right], \tag{16}$$

$$F_2(\infty) = \frac{\phi_2 D_2}{r}. \tag{17}$$

Boundary condition (16) reflects the effect of the seniority of F_1 over F_2 on the payoff to junior debt holders in the event of bankruptcy. From (17), and as in the case of the pricing of senior debt, α_1 must be zero and

$$\alpha_0 = \frac{\phi_2 D_2}{r}. \tag{18}$$

Hence, F_2 becomes

$$F_2 = \frac{\phi_2 D_2}{r} + \alpha_2 S^{-\gamma}. \tag{19}$$

In applying boundary condition (16) to equation (19), three cases must be distinguished:

- *Case 1:* $(1-\alpha)S_B < D_1$.

In this case, the payoff to junior debt holders in the event of bankruptcy is zero, and (16) becomes

$$F_2(S_B) = 0, \tag{20}$$

yielding

$$\alpha_2 = -\frac{\phi_2 D_2}{r} S_B^{\gamma}. \tag{21}$$

Therefore, the value of junior debt, F_2, is given by:

$$F_2 = \frac{\phi_2 D_2}{r}\left(1 - \left(\frac{S}{S_B}\right)^{-\gamma}\right). \tag{22}$$

- *Case 2:* $D_1 < (1-\alpha)S_B < D_1 + D_2$.

Then, boundary condition (16) becomes

$$F_2(S_B) = (1-\alpha)S_B - D_1, \tag{23}$$

implying

$$\alpha_2 = \left((1-\alpha)S_B - D_1 - \frac{\phi_2 D_2}{r} \right) S_B{}^\gamma \tag{24}$$

and

$$F_2 = \frac{\phi_2 D_2}{r} + \left((1-\alpha)S_B - D_1 - \frac{\phi_2 D_2}{r} \right) \left(\frac{S}{S_B} \right)^{-\gamma}. \tag{25}$$

- *Case 3:* $(1-\alpha)S_B > D_1 + D_2$.

In this case, the face amount of junior debt, D_2, is repaid in full in the event of bankruptcy, and (16) becomes

$$F_2(S_B) = D_2, \tag{26}$$

implying

$$\alpha_2 = D_2 \left(1 - \frac{\phi_2}{r} \right) S_B^\gamma, \tag{27}$$

and

$$F_2(S) = \frac{\phi_2 D_2}{r} + D_2 \left(1 - \frac{\phi_2}{r} \right) \left(\frac{S}{S_B} \right)^{-\gamma}. \tag{28}$$

These results can be summarized as follows:

$$F_2(S) = \begin{cases} \dfrac{\phi_2 D_2}{r} \left(1 - \left(\dfrac{S}{S_B} \right)^{-\gamma} \right) & \text{if } (1-\alpha)S_B < D_1 \\[3ex] \dfrac{\phi_2 D_2}{r} + \left((1-\alpha)S_B - D_1 - \dfrac{\phi_2 D_2}{r} \right) \left(\dfrac{S}{S_B} \right)^{-\gamma} & \\[1ex] & \text{if } D_1 < (1-\alpha)S_B < D_1 + D_2 \\[3ex] \dfrac{\phi_2 D_2}{r} + D_2 \left(1 - \dfrac{\phi_2}{r} \right) \left(\dfrac{S}{S_B} \right)^{-\gamma} & \text{if } (1-\alpha)S_B > D_1 + D_2. \end{cases} \tag{29}$$

As was done for senior debt, the value of junior debt (29) can be interpreted in terms of the payoffs to junior bondholders weighted with the risk-neutral probability of bankruptcy, $(S/S_B)^{-\gamma}$.

4.3.3   The Value of the Firm

From Leland (1994), we know that the total value of the firm W reflects three terms: the firm's asset value S, the value of the tax deduction of interest payments TB, less the value of bankruptcy costs K. The value of bankruptcy costs K must satisfy (3) with boundary conditions

$$K(S_B) = \alpha S_B,$$ (30)

$$K(\infty) = 0.$$ (31)

From (31), $\alpha_0 = \alpha_1 = 0$, and from (30), we get

$$K(S_B) = \alpha S_B = \alpha_2 S_B^{-\gamma} \quad \Leftrightarrow \quad \alpha_2 = \alpha S_B^{1+\gamma},$$ (32)

and therefore

$$K(S) = \alpha_2 S^{-\gamma} = \alpha S_B^{1+\gamma} S^{-\gamma}.$$ (33)

Similarly, the value of the tax benefits, TB, must satisfy (3) with boundary conditions

$$TB(S_B) = 0,$$ (34)

$$TB(\infty) = \theta \frac{\phi_1 D_1 + \phi_2 D_2}{r}.$$ (35)

Boundary condition (34) says that the tax benefits are lost if bankruptcy occurs. Boundary condition (35) says that, as the asset value becomes very large and bankruptcy unlikely, the value of the tax benefits approaches the value of the risk-free debt times the tax rate θ. From (34), $\alpha_1 = 0$ and $\alpha_0 = \theta(\phi_1 D_1 + \phi_2 D_2)/r$. Substituting into (3) and using (34), we get

$$TB(S_B) = \theta \frac{\phi_1 D_1 + \phi_2 D_2}{r} + \alpha_2 S_B^{-\gamma} = 0 \quad \Leftrightarrow$$

$$\alpha_2 = -\theta \frac{\phi_1 D_1 + \phi_2 D_2}{r} S_B^{\gamma}$$ (36)

Hence,

$$TB(S) = \theta \frac{\phi_1 D_1 + \phi_2 D_2}{r} \left(1 - \left(\frac{S}{S_B} \right)^{-\gamma} \right).$$ (37)

Using (33) and (37), the total value of the firm, W, equals:

$$W(S) = S + TB(S) - K(S)$$

$$= S + \theta \frac{\phi_1 D_1 + \phi_2 D_2}{r} \left(1 - \left(\frac{S}{S_B} \right)^{-\gamma} \right) - \alpha S_B^{1+\gamma} S^{-\gamma}.$$ (38)

As in Chapter 3, the total value of the firm, W, equals current asset value, S, plus the present value of the tax shields, $\theta(\phi_1 D_1 + \phi_2 D_2)/r$, times the risk-neutral probability that bankruptcy does not occur, $1 - (S/S_B)^{-\gamma^*}$, minus the value lost in the event of bankruptcy, αS_B, times the risk-neutral probability of bankruptcy, $(S/S_B)^{-\gamma^*}$.

4.3.4   The Value of Equity

The value of equity, E, is the total value of the firm W less the value of outstanding debt, $F_1 + F_2$:

$$E(S) = W(S) - \left(F_1(S) + F_2(S)\right). \tag{39}$$

Because the bankruptcy strategy of the equity holders is still unknown, however, three cases have to be distinguished:

- *Case 1:* $(1-\alpha)S_B < D_1$.

In this case,

$$F_1(S) = \frac{\phi_1 D_1}{r} + \left((1-\alpha)S_B - \frac{\phi_1 D_1}{r}\right)\left(\frac{S}{S_B}\right)^{-\gamma} \tag{10}$$

and

$$F_2 = \frac{\phi_2 D_2}{r}\left(1 - \left(\frac{S}{S_B}\right)^{-\gamma}\right). \tag{22}$$

Hence,

$$E(S) = S + \frac{\theta(\phi_1 D_1 + \phi_2 D_2)}{r}\left(1 - \left(\frac{S}{S_B}\right)^{-\gamma}\right) - \alpha S_B^{1+\gamma} S^{-\gamma}$$

$$- \left(\frac{\phi_1 D_1}{r} + \left((1-\alpha)S_B - \frac{\phi_1 D_1}{r}\right)\left(\frac{S}{S_B}\right)^{-\gamma} + \frac{\phi_2 D_2}{r}\left(1 - \left(\frac{S}{S_B}\right)^{-\gamma}\right)\right) \tag{40}$$

$$= S - \frac{(1-\theta)(\phi_1 D_1 + \phi_2 D_2)}{r}\left(1 - \left(\frac{S}{S_B}\right)^{-\gamma}\right) - S_B^{1+\gamma} S^{-\gamma}.$$

- *Case 2:* $D_1 < (1-\alpha)S_B < D_1 + D_2$.

Then,

$$F_1(S) = \frac{\phi_1 D_1}{r} + D_1\left(1 - \frac{\phi_1}{r}\right)\left(\frac{S}{S_B}\right)^{-\gamma} \tag{13}$$

and

$$F_2 = \frac{\phi_2 D_2}{r} + \left((1-\alpha)S_B - D_1 - \frac{\phi_2 D_2}{r}\right)\left(\frac{S}{S_B}\right)^{-\gamma}. \tag{25}$$

Therefore,

$$E(S) = S + \frac{\theta(\phi_1 D_1 + \phi_2 D_2)}{r}\left(1 - \left(\frac{S}{S_B}\right)^{-\gamma}\right) - \alpha S_B^{1+\gamma} S^{-\gamma}$$

$$- \left(\frac{\phi_1 D_1}{r} + D_1\left(1 - \frac{\phi_1}{r}\right)\left(\frac{S}{S_B}\right)^{-\gamma} + \frac{\phi_2 D_2}{r}\right.$$

$$\left. + \left((1-\alpha)S_B - D_1 - \frac{\phi_2 D_2}{r}\right)\left(\frac{S}{S_B}\right)^{-\gamma}\right)$$

$$(41)$$

$$= S - \frac{(1-\theta)(\phi_1 D_1 + \phi_2 D_2)}{r}\left(1 - \left(\frac{S}{S_B}\right)^{-\gamma}\right) - S_B^{1+\gamma} S^{-\gamma}.$$

- *Case 3:* $(1-\alpha)S_B > D_1 + D_2$.

Then,

$$F_1(S) = \frac{\phi_1 D_1}{r} + D_1\left(1 - \frac{\phi_1}{r}\right)\left(\frac{S}{S_B}\right)^{-\gamma} \qquad (13)$$

and

$$F_2(S) = \frac{\phi_2 D_2}{r} + D_2\left(1 - \frac{\phi_2}{r}\right)\left(\frac{S}{S_B}\right)^{-\gamma}. \qquad (28)$$

Therefore,

$$E(S) = S + \frac{\theta(\phi_1 D_1 + \phi_2 D_2)}{r}\left(1 - \left(\frac{S}{S_B}\right)^{-\gamma}\right) - \alpha S_B^{1+\gamma} S^{-\gamma}$$

$$- \left(\frac{\phi_1 D_1}{r} + D_1\left(1 - \frac{\phi_1}{r}\right)\left(\frac{S}{S_B}\right)^{-\gamma} + \frac{\phi_2 D_2}{r} + D_2\left(1 - \frac{\phi_2}{r}\right)\left(\frac{S}{S_B}\right)^{-\gamma}\right)$$

$$(42)$$

$$= S - \frac{(1-\theta)(\phi_1 D_1 + \phi_2 D_2)}{r}\left(1 - \left(\frac{S}{S_B}\right)^{-\gamma}\right)$$

$$- (D_1 + D_2)\left(\frac{S}{S_B}\right)^{-\gamma} - \alpha S_B^{1+\gamma} S^{-\gamma}.$$

These results can be summarized as follows:

$$E(S) = \begin{cases} S - \dfrac{(1-\theta)(\phi_1 D_1 + \phi_2 D_2)}{r}\left(1 - \left(\dfrac{S}{S_B}\right)^{-\gamma}\right) - S_B^{1+\gamma} S^{-\gamma} \\ \qquad\qquad\qquad\qquad if \quad (1-\alpha)S_B < D_1 + D_2 \\[2ex] S - \dfrac{(1-\theta)(\phi_1 D_1 + \phi_2 D_2)}{r}\left(1 - \left(\dfrac{S}{S_B}\right)^{-\gamma}\right) \\[2ex] \qquad - (D_1 + D_2)\left(\dfrac{S}{S_B}\right)^{-\gamma} - \alpha S_B^{1+\gamma} S^{-\gamma} \\ \qquad\qquad\qquad\qquad if \quad (1-\alpha)S_B > D_1 + D_2. \end{cases} \tag{43}$$

Figure 4.2 illustrates the results of this section by plotting the value of the firm and its securities for different values of the bankruptcy trigger S_B, which is still arbitrary at this point. As was the case in Figure 3.2, firm value falls as the bankruptcy trigger S_B is increased. Equity value E rises at first, and then falls. Finally, the value of both senior debt F_1 and junior debt F_2 falls, and then rises, but junior debt is far more sensitive to changes in S_B than senior debt.

Now that the players' payoffs have been valued, the game can be solved for the players' optimal strategies, starting with the last decision, namely, the equity holders' bankruptcy decision.

4.4 The Equity Holders' Optimal Bankruptcy Choice

Using equation (43), it is now possible to determine the bankruptcy-triggering asset value S_B. It is chosen by the equity holders so as to maximize the current value of equity. In the case $(1-\alpha)S_B < D_1 + D_2$, this maximization leads to the first-order condition

$$\frac{\partial E(S)}{\partial S_B} = \frac{(1-\theta)(\phi_1 D_1 + \phi_2 D_2)}{r}\gamma S_B^{\gamma-1} S^{-\gamma} - (1+\gamma)S_B^{\gamma} S^{-\gamma} = 0, \tag{44}$$

yielding[1]

---

[1] It is a maximum, since

$$\frac{\partial^2 E(S)}{\partial S_B^2} = (1-\theta)\frac{(\phi_1 D_1 + \phi_2 D_2)}{r}\gamma(\gamma-1)S_B^{\gamma-2}S^{-\gamma} - (1+\gamma)\gamma S_B^{\gamma-1}S^{-\gamma}$$

$$= S_B^{\gamma-2}S^{-\gamma}\gamma\left((1-\theta)\frac{(\phi_1 D_1 + \phi_2 D_2)}{r}(\gamma-1) - (1+\gamma)S_B\right).$$

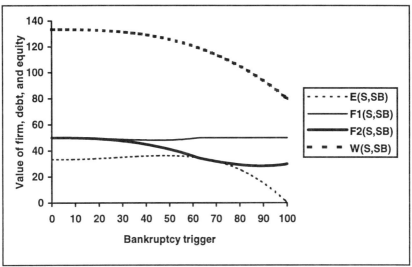

Figure 4.2: *Value of the firm, debt and equity as a function of the bankruptcy-triggering asset level* S_B *for the following parameter values:* $\theta = 1/3$, $\alpha = 0.2$, $S = 100$, $r = \phi_1 = \phi_2 = 0.05$, $D_1 = D_2 = 50$ *and* $\sigma = 0.2$. *As* S_B *is increased, firm value W falls. Equity value E rises at first, and then falls. Finally, the value of both senior debt* F_1 *and junior debt* F_2 *falls, and then rises, but junior debt is far more sensitive to changes in* S_B *than senior debt.*

$$S_B = \frac{(1-\theta)(\phi_1 D_1 + \phi_2 D_2)}{r + \sigma^2/2}. \tag{45}$$

Notice that this value doesn't depend on current asset value. S. The question of whether $(1-\alpha)S_B < D_1$ or $(1-\alpha)S_B > D_1$ will depend on the parameters in (45) and cannot be answered in general.

In the case where $(1-\alpha)S_B > D_1 + D_2$, the first-order condition is

$$\frac{\partial E(S)}{\partial S_B} = \left(\frac{(1-\theta)(\phi_1 D_1 + \phi_2 D_2)}{r} - (D_1 + D_2) \right) \gamma S_B^{\gamma-1} S^{-\gamma}$$

$$- \alpha(1+\gamma)S_B^{\gamma} S^{-\gamma} = 0 \tag{46}$$

$$= S_B^{\gamma-2} S^{-\gamma} \gamma(1-\theta) \left(\frac{(\phi_1 D_1 + \phi_2 D_2)}{r}(\gamma - 1) - (1+\gamma)\frac{(\phi_1 D_1 + \phi_2 D_2)\gamma}{r(1+\gamma)} \right)$$

$$= -S_B^{\gamma-2} S^{-\gamma} \gamma(1-\theta)\frac{(\phi_1 D_1 + \phi_2 D_2)}{r} < 0.$$

and the optimal value for S_B equals:[2]

$$S_B = \frac{\gamma}{\alpha(\gamma+1)}\left(\frac{(1-\theta)(\phi_1 D_1 + \phi_2 D_2)}{r} - (D_1 + D_2)\right). \qquad (47)$$

The question of whether (45) or (47) is the optimal bankruptcy strategy cannot be answered generally, since the corresponding equity values will depend on the model parameters. In the sequel, we assume that (45) is optimal. This arbitrary choice allows us to analyze the interesting case in which the lender defaults partially on (at least) the junior claim.[3]

---

[2] It is a maximum, since

$$\frac{\partial^2 E(S)}{\partial S_B^2} = \left(\frac{(1-\theta)(\phi_1 D_1 + \phi_2 D_2)}{r} - (D_1 + D_2)\right)\gamma(\gamma-1)S_B^{\gamma-2}S^{-\gamma} - \alpha\gamma(1+\gamma)S_B^{\gamma-1}S^{-\gamma}$$

$$= \gamma S_B^{\gamma-2}S^{-\gamma}\left(\left(\frac{(1-\theta)(\phi_1 D_1 + \phi_2 D_2)}{r} - (D_1 + D_2)\right)(\gamma-1) - \alpha(1+\gamma)S_B\right)$$

$$= -\gamma S_B^{\gamma-2}S^{-\gamma}\left(\frac{(1-\theta)(\phi_1 D_1 + \phi_2 D_2)}{r} - (D_1 + D_2)\right) < 0.$$

[3] This assumption need not hold. To see this, notice first that the bankruptcy strategy $S_B = (D_1 + D_2)/(1-\alpha)$ cannot be optimal:

$$\left.\frac{\partial E(S)}{\partial S_B}\right|_{S_B \uparrow \frac{D_1+D_2}{1-\alpha}} = \left(\frac{D_1+D_2}{(1-\alpha)S}\right)^\gamma\left((1-\theta)(1-\alpha)\gamma\frac{\phi_1 D_1 + \phi_2 D_2}{r(D_1 + D_2)} - (1+\gamma)\right)$$

together with

$$\left.\frac{\partial E(S)}{\partial S_B}\right|_{S_B \downarrow \frac{D_1+D_2}{1-\alpha}} = \left(\frac{D_1+D_2}{(1-\alpha)S}\right)^\gamma\left((1-\theta)(1-\alpha)\gamma\frac{\phi_1 D_1 + \phi_2 D_2}{r(D_1 + D_2)} - (\alpha+\gamma)\right)$$

imply

$$\left.\frac{\partial E(S)}{\partial S_B}\right|_{S_B \uparrow \frac{D_1+D_2}{1-\alpha}} < \left.\frac{\partial E(S)}{\partial S_B}\right|_{S_B \downarrow \frac{D_1+D_2}{1-\alpha}}.$$

Now suppose that $\left.\partial E(S)/\partial S_B\right|_{S_B \uparrow \frac{D_1+D_2}{1-\alpha}} < 0$. Then there exists a bankruptcy trigger $\overline{S}_B < (D_1 + D_2)/(1-\alpha)$ such that $E(S_B = \overline{S}_B) > E(S_B = (D_1 + D_2)/(1-\alpha))$. Alternatively, if $\left.\partial E(S)/\partial S_B\right|_{S_B \uparrow \frac{D_1+D_2}{1-\alpha}} \geq 0$, then $\left.\partial E(S)/\partial S_B\right|_{S_B \downarrow \frac{D_1+D_2}{1-\alpha}} > 0$ and there must exist a bankruptcy trigger $\overline{S}_B > (D_1 + D_2)/(1-\alpha)$ with the property that $E(S_B = \overline{S}_B) > E(S_B = (D_1 + D_2)/(1-\alpha))$. Hence, $S_B = (D_1 + D_2)/(1-\alpha)$ cannot be optimal. It follows then that a sufficient condition for (45) not to be optimal is

4.5 The Firm's Decision to Issue Junior Claims

Once the bankruptcy decision has been analyzed, the next question to be answered when solving the game is that of knowing if and when the firm will decide to issue junior claims. First, however, we wish to know when it will be *socially* optimal to increase leverage. Such will be the case if

$$\left.\frac{\partial W(S)}{\partial (\phi_2 D_2)}\right|_{\phi_2 D_2 = 0} > 0, \tag{48}$$

that is, if issuing junior debt raises the total value of the firm. Differentiating (38) partially with respect to $\phi_2 D_2$ yields

$$\frac{\partial W(S)}{\partial (\phi_2 D_2)} = \frac{\theta}{r}\left(1 - \left(\frac{S}{S_B}\right)^{-\gamma}\right) - \theta\frac{\phi_1 D_1 + \phi_2 D_2}{r}\frac{\gamma}{S_B}\frac{1-\theta}{r+\sigma^2/2}\left(\frac{S}{S_B}\right)^{-\gamma}$$

$$- \alpha(1+\gamma)\frac{1-\theta}{r+\sigma^2/2}S_B^\gamma S^{-\gamma} \tag{49}$$

$$= \frac{\theta}{r} - \left(\frac{\theta}{r} + \frac{\theta + \alpha(1-\theta)}{\sigma^2/2}\right)\left(\frac{S}{S_B}\right)^{-\gamma}.$$

This expression will be positive if and only if

$$S > \frac{(1-\theta)\phi_1 D_1}{r+\sigma^2/2}\left(\frac{\theta + \gamma\big(\theta + \alpha(1-\theta)\big)}{\theta}\right)^{1/\gamma}, \tag{50}$$

that is, if current asset value is sufficiently high.

Equation (50) gives the necessary and sufficient condition for the issue of junior debt to be socially optimal. But when is such an issue likely to be made? To answer this question, some additional assumptions on what the equity holders will do with the amount collected in the issue are required. Suppose that this amount is paid out to shareholders. Then, shareholders seek to maximize the amount they are entitled to, which is equal to the value of equity, E, plus the amount paid out, F_2:

$$E(S) + F_2(S) = W(S) - F_1(S). \tag{51}$$

$$\left.\frac{\partial E(S)}{\partial S_B}\right|_{S_B \uparrow \frac{D_1 + D_2}{1-\alpha}} = \left(\frac{D_1 + D_2}{(1-\alpha)S}\right)^\gamma\left((1-\theta)(1-\alpha)\gamma\frac{\phi_1 D_1 + \phi_2 D_2}{r(D_1 + D_2)} - (1+\gamma)\right) > 0,$$

or

$$\frac{\phi_1 D_1 + \phi_2 D_2}{r(D_1 + D_2)} > \frac{1+\gamma}{\gamma(1-\theta)(1-\alpha)}.$$

The problem with equation (51) is that its value depends on the bankruptcy decision of the shareholders and that we do not know whether it is such that $(1-\alpha)S_B < D_1$ or $(1-\alpha)S_B > D_1$. Therefore, two cases have to be distinguished:

- *Case 1:* $(1-\alpha)S_B > D_1$

Suppose first that $(1-\alpha)S_B > D_1$. Then, this condition will be satisfied as well after the additional debt has been issued. To see this, remember that $S_B = (1-\theta)(\phi_1 D_1 + \phi_2 D_2)/(r+\sigma^2/2)$. Then it is clear that S_B rises when $\phi_2 D_2$ increases, that is, the bankruptcy-triggering asset value increases when junior debt is issued.

Remember that when $(1-\alpha)S_B > D_1$, the value of senior debt is given by

$$F_1(S) = \frac{\phi_1 D_1}{r} + D_1\left(1-\frac{\phi_1}{r}\right)\left(\frac{S}{S_B}\right)^{-\gamma}. \tag{13}$$

Then,

$$
\begin{aligned}
W(S) - F_1(S) &= S + \theta\frac{\phi_1 D_1 + \phi_2 D_2}{r}\left(1-\left(\frac{S}{S_B}\right)^{-\gamma}\right) - \alpha S_B^{1+\gamma} S^{-\gamma} \\
&\quad - \left(\frac{\phi_1 D_1}{r}\left(1-\left(\frac{S}{S_B}\right)^{-\gamma}\right) + D_1\left(\frac{S}{S_B}\right)^{-\gamma}\right) \\
&= S + \frac{(\theta-1)\phi_1 D_1 + \theta\phi_2 D_2}{r}\left(1-\left(\frac{S}{S_B}\right)^{-\gamma}\right) \\
&\quad - D_1\left(\frac{S}{S_B}\right)^{-\gamma} - \alpha S_B^{1+\gamma} S^{-\gamma}.
\end{aligned} \tag{52}
$$

Therefore, we have

$$
\begin{aligned}
\frac{\partial(W(S) - F_1(S))}{\partial(\phi_2 D_2)} &= \frac{\theta}{r}\left(1-\left(\frac{S}{S_B}\right)^{-\gamma}\right) - \alpha(1+\gamma)\frac{1-\theta}{r+\sigma^2/2}\left(\frac{S}{S_B}\right)^{-\gamma} \\
&\quad - \frac{\gamma}{S_B}\frac{1-\theta}{r+\sigma^2/2}\left(\frac{(\theta-1)\phi_1 D_1 + \theta\phi_2 D_2}{r} + D_1\right)\left(\frac{S}{S_B}\right)^{-\gamma} \\
&= \frac{1}{r}\left(\theta - \left(\frac{S}{S_B}\right)^{-\gamma}\left(\theta + \gamma(\theta+\alpha(1-\theta)) + \frac{\gamma D_1(r-\phi_1)}{\phi_1 D_1 + \phi_2 D_2}\right)\right).
\end{aligned} \tag{53}
$$

For this expression to be positive, we must have

$$\left(\frac{S}{S_B}\right)^{\gamma} > \frac{\theta + \gamma(\theta + \alpha(1-\theta)) + \dfrac{\gamma D_1(r-\phi_1)}{\phi_1 D_1 + \phi_2 D_2}}{\theta}, \tag{54}$$

or, evaluating this expression at $\phi_2 D_2 = 0$:

$$S > \frac{(1-\theta)\phi_1 D_1}{r + \sigma^2/2} \left(\frac{\theta + \gamma(\theta + \alpha(1-\theta)) + \gamma\left(\dfrac{r}{\phi_1} - 1\right)}{\theta}\right)^{1/\gamma}. \tag{55}$$

Comparing condition (55) with (50), we see that, for a junior debt issue to be profitable to the borrower, S must be *higher* than in the social optimum if $r > \phi_1$ and *lower* if $r < \phi_1$. The reason is that, by triggering bankruptcy earlier, the issue increases the value of senior debt whenever $r > \phi_1$, as inspection of equation (13) shows. In other words, with $r > \phi_1$, a part of the social gain achieved through a capital structure change is reaped by senior debt holders. If $r < \phi_1$, the opposite is true. In the special case where $r = \phi_1$, the value of senior debt equals that of a risk-free asset:

$$F_1(S) = \frac{\phi_1 D_1}{r} + D_1\left(1 - \frac{\phi_1}{r}\right)\left(\frac{S}{S_B}\right)^{-\gamma} = \frac{\phi_1 D_1}{r} = D_1. \tag{56}$$

Therefore, the issue of junior debt has no consequences on the value of senior debt.[4]

• *Case 2:* $(1-\alpha)S_B < D_1$

Now consider the situation in which $(1-\alpha)S_B < D_1$. Then, using the expression

$$F_1(S) = \frac{\phi_1 D_1}{r} + \left((1-\alpha)S_B - \frac{\phi_1 D_1}{r}\right)\left(\frac{S}{S_B}\right)^{-\gamma}, \tag{10}$$

we can compute the equity holders' payoff, $E(S) + F_2(S) = W(S) - F_1(S)$, as

---

[4] We will return to this point in Section 4.6 and demonstrate that (56) is, in fact, never satisfied.

$$W(S) - F_1(S) = S + \theta \frac{\phi_1 D_1 + \phi_2 D_2}{r}\left(1 - \left(\frac{S}{S_B}\right)^{-\gamma}\right) - \alpha S_B^{1+\gamma} S^{-\gamma}$$

$$-\left(\frac{\phi_1 D_1}{r}\left(1 - \left(\frac{S}{S_B}\right)^{-\gamma}\right) + (1-\alpha)S_B\left(\frac{S}{S_B}\right)^{-\gamma}\right)$$

(57)

$$= S + \theta \frac{\phi_1 D_1 + \phi_2 D_2}{r}\left(1 - \left(\frac{S}{S_B}\right)^{-\gamma}\right)$$

$$- \frac{\phi_1 D_1}{r}\left(1 - \left(\frac{S}{S_B}\right)^{-\gamma}\right) - S_B^{1+\gamma} S^{-\gamma}.$$

Differentiating this expression partially with respect to $\phi_2 D_2$ yields

$$\frac{\partial(W(S) - F_1(S))}{\partial(\phi_2 D_2)} = \frac{\theta}{r}\left(1 - \left(\frac{S}{S_B}\right)^{-\gamma}\right)$$

$$- \theta \frac{\phi_1 D_1 + \phi_2 D_2}{r}\frac{\gamma}{S_B}\frac{1-\theta}{r+\sigma^2/2}\left(\frac{S}{S_B}\right)^{-\gamma}$$

(58)

$$+ \frac{\phi_1 D_1}{r}\frac{\gamma}{S_B}\frac{1-\theta}{r+\sigma^2/2}\left(\frac{S}{S_B}\right)^{-\gamma} - (1+\gamma)\frac{1-\theta}{r+\sigma^2/2}\left(\frac{S}{S_B}\right)^{-\gamma}$$

$$= \frac{1}{r}\left(\theta - \left(\frac{S}{S_B}\right)^{-\gamma}\left(\theta + \gamma\left(1 - \frac{\phi_1 D_1}{\phi_1 D_1 + \phi_2 D_2}\right)\right)\right).$$

For this expression to be positive at $\phi_2 D_2 = 0$, we must have

$$S > S_B = \frac{(1-\theta)\phi_1 D_1}{r+\sigma^2/2},$$

(59)

which means that a junior debt issue makes sense as long as the bankruptcy trigger hasn't been reached. Thus, condition (59) is merely the tendency that equity holders have to increase payouts, which has been illustrated in Section 3.8.3, in another shape.

What are the consequences of these results for capital structure? The equity holders' decision to issue junior debt is governed by equation (55) or (59), which do not match the condition required for a junior debt issue to be socially optimal (equation (50)). This means that the existence of senior debt *distorts* the equity holders' decision to issue junior debt, thus leading to socially suboptimal capital structures, as Figure 4.3 illustrates.

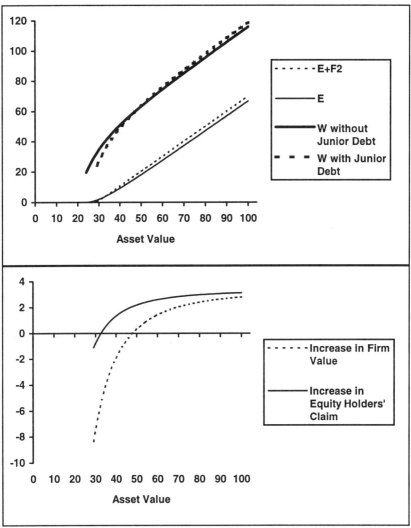

Figure 4.3: *Effect of a junior debt issue on the value of the equity holders' claim, $E + F_2$, and firm value W as a function of current asset value S for the following parameter values: $\theta = 1/3$, $\alpha = 0.2$, $r = \phi_1 = \phi_2 = 0.05$, $D_1 = 50$, $D_2 = 10$ and $\sigma = 0.2$. Over some range of asset values, a junior debt issue increases the value of the equity holders' claim, but reduces overall firm value, thus giving rise to a principal-agent problem.*

The reason for this result is that issuing junior debt transfers wealth away from senior creditors to equity holders. We will now analyze this point in more detail, because it demonstrates that seniority does *not* protect debt holders against adverse wealth effects resulting from subsequent debt issues.

4.6 The Influence of Junior Debt on the Value of Senior Debt

The previous section showed that the equity holders' decision to issue junior debt is distorted by the existence of senior debt and argued that the reason for this result lies in the influence of junior debt issues on the value of senior debt. In this section, we demonstrate that no senior debt contract exists that *immunizes* senior debt holders against the issue of junior debt. In Section 4.6.1, we first show that no contract can be found that immunizes senior debt holders altogether. Section 4.6.2 will then be concerned with immunization against *negative* wealth effects.

4.6.1 On the Impossibility of Total Immunization

To analyze the effect of junior debt issues on the value of senior debt, let us compute the partial derivative of the value senior debt with respect to $\phi_2 D_2$. If $(1-\alpha)S_B < D_1$, then

$$\frac{\partial F_1(S)}{\partial(\phi_2 D_2)} = -\frac{\phi_1 D_1}{r}\frac{\gamma}{S_B}\frac{1-\theta}{r+\sigma^2/2}\left(\frac{S}{S_B}\right)^{-\gamma}$$

$$+(1-\alpha)(1+\gamma)\frac{1-\theta}{r+\sigma^2/2}\left(\frac{S}{S_B}\right)^{-\gamma} \qquad (60)$$

$$= \frac{1}{\sigma^2/2}\left(\frac{S}{S_B}\right)^{-\gamma}\left((1-\alpha)(1-\theta)-\frac{\phi_1 D_1}{\phi_1 D_1 + \phi_2 D_2}\right)$$

Evaluating this expression at $\phi_2 D_2 = 0$ yields

$$\frac{\partial F_1(S)}{\partial(\phi_2 D_2)} = \frac{1}{\sigma^2/2}\left(\frac{S}{S_B}\right)^{-\gamma}\left((1-\alpha)(1-\theta)-1\right)<0, \qquad (61)$$

and it is clear that a junior debt issue has a negative influence on the value of senior debt.

Now consider the case in which $(1-\alpha)S_B > D_1$. Then,

$$F_1(S) = \frac{\phi_1 D_1}{r} + D_1\left(1-\frac{\phi_1}{r}\right)\left(\frac{S}{S_B}\right)^{-\gamma}, \qquad (13)$$

and therefore,

$$\frac{\partial F_1(S)}{\partial(\phi_2 D_2)} = D_1 \left(1 - \frac{\phi_1}{r}\right)\left(\frac{S}{S_B}\right)^{-\gamma} \frac{\gamma}{S_B} \frac{1-\theta}{r+\sigma^2/2}$$
$$= \frac{2D_1(r-\phi_1)}{\sigma^2(\phi_1 D_1 + \phi_2 D_2)}\left(\frac{S}{S_B}\right)^{-\gamma}.$$
(62)

Now, senior debt holders could immunize against a junior debt issue by choosing $\phi_1 = r$. But then, from the analysis in Section 4.4 above and from the fact that $\phi_2 D_2 = 0$, the equity holders' optimal bankruptcy choice is given by

$$S_B = \frac{(1-\theta)(\phi_1 D_1 + \phi_2 D_2)}{r+\sigma^2/2} = \frac{(1-\theta)\phi_1 D_1}{r+\sigma^2/2} < D_1,$$
(63)

which contradicts our assumption that $(1-\alpha)S_B > D_1$. Hence, in spite of senior debt having priority over other claims, its value is influenced by the issue of junior claims. There is no way in which senior debt can be immunized against the issue of junior claims.

4.6.2 On the Impossibility of Immunization against Negative Influence

In practice, however, senior debt holders will be interested in avoiding a *negative* influence of a junior debt issue on the value of their claim. Is this aim achievable? From equation (61), it is clear that this is impossible in the case where $(1-\alpha)S_B < D_1$. In the case where $(1-\alpha)S_B > D_1$, we would require

$$\frac{\partial F_1(S)}{\partial(\phi_2 D_2)} = \frac{2D_1(r-\phi_1)}{\sigma^2(\phi_1 D_1 + \phi_2 D_2)}\left(\frac{S}{S_B}\right)^{-\gamma} > 0,$$
(64)

that is

$$r - \phi_1 > 0.$$
(65)

But (65) implies $(1-\alpha)S_B < D_1$, a contradiction. Therefore, immunization against the adverse effects of a junior debt issue is impossible as well. Figure 4.4 gives an example of the effect of a junior debt issue on the value of senior debt. When junior debt with a face value of 10 is issued, the value of senior debt falls. This illustrates the above result that seniority need not protect debt holders against adverse wealth changes resulting from the issue of new debt. This wealth transfer from senior debt holders to equity holders distorts the equity holders' decision to issue junior debt and gives rise to suboptimal capital structures.

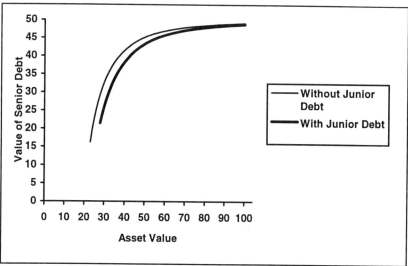

Figure 4.4: *Effect of a junior debt issue on the value of senior debt as a function of current asset value S for the following parameter values:* $\theta = 1/3$, $\alpha = 0.2$, $r = \phi_1 = \phi_2 = 0.05$, $D_1 = 50$, $D_2 = 10$ *and* $\sigma = 0.2$. *When junior debt with a face value of 10 is issued, the value of senior debt falls. Through the issue, shareholders transfer wealth away from senior bondholders. Thus, seniority need not protect senior debt holders against adverse wealth changes resulting from the issue of new debt.*

4.7 Conclusion

The analysis presented in this chapter has shown how senior and junior debt could be priced by taking the seniority of claims into account in the boundary conditions of the contingent claims pricing formula. After valuing senior and junior debt, the firm, and equity, it was shown that when bankruptcy is endogenous, the existence of junior debt influences the equity holders' optimal bankruptcy choice and therefore the timing of bankruptcy.

Moreover, it was demonstrated that the existence of senior debt *distorts* the equity holders' decision to issue junior debt. The reason is that a change in capital structure involving the issue of junior debt and the payout of the proceeds to equity holders reduces the market value of senior debt. The existence of this wealth transfer from senior debt holders to equity holders invalidates the conventional wisdom that seniority protects debt holders against adverse wealth changes resulting from the issue of new debt. It was demonstrated that there exists no contract immunizing senior debt holders from the negative influence of a junior debt issue on the value of their claim. The reason for this result is that the existence of junior debt changes the firm's bankruptcy decision and therefore the timing of bankruptcy.

Besides these theoretical considerations, the distortion of the equity holders' decision to issue junior debt presented in this chapter implies that socially suboptimal capital structures are likely to arise in practice *in spite of seniority*. In other words, seniority does not resolve the principal-agent problem of junior debt issues.

5. Bank Runs

5.1 Introduction

Bank Runs are one of the most puzzling phenomena of banking history. Calomiris (1997) reports that banking crises in ancient Greece and Rome date from at least the 4th century B.C., as do government interventions to alleviate them.

Diamond and Dybvig (1983) depict how banks can provide risk-sharing potential to depositors unaware of their future consumption needs. They show that this risk-sharing function provides both the rationale for the existence of banks and for their vulnerability to runs, even if there is no uncertainty over the value of the projects the bank has invested in.

This chapter considers bank runs in a world in which the payoff of the projects in which the bank has invested may be uncertain. In a continuous-time framework, we show that banks are indeed vulnerable to runs. Moreover, the possibility of bank runs has a very important implication for the role of financial intermediaries as providers of *capital*.

The focus of the chapter, however, lies on the *incentive effects* of the possibility of bank runs. The analysis provided below demonstrates that the existence of bank runs disciplines bank behavior. The reason is that banks anticipate the possibility of runs and reduce the risk of their investments so as to avoid them. Therefore, in the simple setting presented below, runs would not occur in equilibrium.

5.2 The Model

Consider a bank with two depositors A and B, each having a deposit of X_0 dollars at initial time 0. Suppose that the bank invests this money in a risky asset with initial price $S_0 = X_0$ and whose value S follows a geometric Brownian motion

$$dS = \mu S dt + \sigma S d\tilde{z} . \tag{1}$$

Assume that a continuous interest rate of $r*$ is paid on the deposits and that the risk-free rate of interest is r, where $r* < r$. Hence, each depositor's claim at time t is given by

$$X(t) = X_0 \cdot e^{r*t} . \tag{2}$$

Furthermore, assume that the depositors are allowed to withdraw the full amount of their deposit at any time without prior notice. If the bank has to liquidate the project it has invested in, it must, however, incur a proportional cost of α. Note that in this setting, financial intermediation is viable if the money remains deposited long enough at the bank so that the interest differential (or *deposit spread*) $r - r^*$ earned by the bank makes up the expected liquidation cost of α.

The analysis in this chapter abstracts from possible conflicts of interest between bank management and shareholders. Throughout, it is assumed that bank management makes decisions that maximize the value of equity.

The structure of the game between bank and depositors is depicted in Figure 5.1. First, bank equity holders decide on how much capital to provide to the bank and depositors decide whether to deposit their money with the bank or not. In a second phase, the bank chooses its investment strategy. If a run seems imminent, bank equity holders must decide on whether to recapitalize the bank or not. Finally, depositors decide on whether to run on the bank or not. If they decide to run, bank assets are liquidated and payoffs are realized.

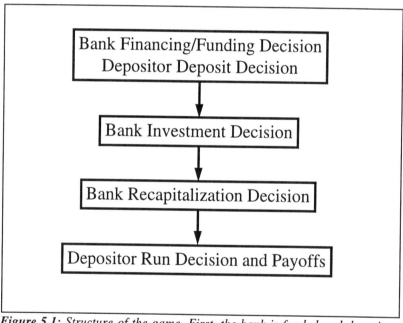

Figure 5.1: *Structure of the game. First, the bank is funded and depositors decide whether to deposit their money or not. The bank then chooses an investment strategy. If a run seems imminent, bank equity holders choose whether to recapitalize the bank or not. Finally, depositors might decide to run on the bank. If they do so, bank assets are liquidated and payoffs are realized.*

The remainder of the chapter is organized as follows: Section 5.3 analyzes the depositors' decision of whether to run on the bank or not. Section 5.4

describes the function of banks as providers of capital. Section 5.5 values the bank's equity as a knock-out call option. Section 5.6 explores the equity holders' decision to recapitalize the bank and shows that recapitalization can be expected to occur when asset value falls below a certain level. Section 5.7 determines the bank's optimal investment decision when bank runs are possible and shows that the possibility of runs disciplines bank behavior. Section 5.8 discusses consequences of the possibility of runs for the bank's funding policy. Section 5.9 comments on the equilibrium deposit spread and the role of transactions costs for the viability of financial intermediation. Section 5.10 concludes the chapter.

5.3 The Incentive to Start a Run

The first question to be answered when solving the game depicted in Figure 5.1 above is whether bank runs will, under the assumptions of Section 5.2, be likely to occur, and if yes, when. To answer this question, let a depositor's withdrawal decision be modeled as a two-person game. Instead of writing down the payoff matrix of the game and searching for the Nash equilibria, simply consider the payoff to the players. The payoff to each of the players, which is depicted in Figure 5.2 below, is given by

$$Min[2(1-\alpha)S_t; X(t)] \qquad (3)$$

if he withdraws first, and

$$Max[0; 2(1-\alpha)S_t - X(t)] \qquad (4)$$

if he doesn't.

When could a run occur? Clearly, the bank is vulnerable to a run as soon as the situation arises in which both depositors want to withdraw first, that is, as soon as

$$Min[2(1-\alpha)S_t; X(t)] > Max[0; 2(1-\alpha)S_t - X(t)]. \qquad (5)$$

From Figure 5.2 below, it is immediately apparent that this will be the case if and only if

$$S_t < \frac{X(t)}{1-\alpha}. \qquad (6)$$

Condition (6) means that a run is possible as soon as the value of bank assets after liquidation is lower than the face value of deposits: $(1-\alpha)S_t < X(t)$.

5.4 A Preliminary Condition for the Existence of Banks

As the previous analysis has shown, depositors have an incentive to run on the bank as soon as $S_t < X(t)/(1-\alpha)$. What are the implications of this fact for the bank?

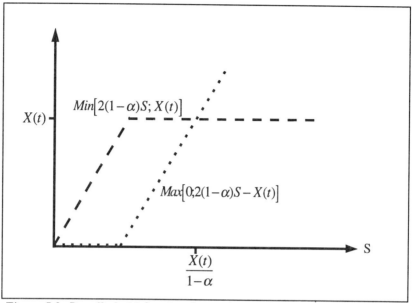

Figure 5.2: *Payoff of withdrawing first and second for different asset values. If current asset value is lower than $X(t)/(1-\alpha)$, depositors have an incentive to try to withdraw first, thus triggering a bank run.*

The most obvious is that, at initial time, when $S_0 = X_0$, the bank might be subject to a run, that is, the depositors will *not* deposit their money at the bank. As a result, the bank must find a way to induce potential depositors to deposit. But how?

The simplest way for the bank to induce this is to provide some capital at initial time. To see this, suppose the bank owners agree to add x dollars of equity for each dollar deposited at the bank. Then, by analogy with the analysis above, *no* run will occur as long as

$$(1+x)S_t > \frac{X(t)}{1-\alpha} \quad \Leftrightarrow \quad S_t > \frac{X(t)}{(1-\alpha)(1+x)} \equiv \overline{S}_t . \qquad (7)$$

Potential depositors can be induced to deposit their money at the bank if condition (7) holds at initial time. Substituting $S_0 = X_0$ in this last condition yields

$$X_0 > \frac{X_0}{(1-\alpha)(1+x)} \quad \Leftrightarrow \quad (1-\alpha)(1+x) > 1 \quad \Leftrightarrow \quad x > \frac{\alpha}{1-\alpha}. \qquad (8)$$

Hence, depositors will agree to deposit their money if the bank or, more precisely, its shareholders agree to compensate for expected liquidation costs by providing $\alpha/(1-\alpha)$ dollars of *capital* for each dollar deposited.

5.5 Valuing the Bank's Equity

After having stressed that a bank can only exist if it provides capital at initial time, the bank's equity can be valued. The aim of this section is to provide an analytical valuation formula for this equity, which will then allow us to analyze incentive problems on the part of the bank. To facilitate this valuation, we make the following additional assumptions:

Assumption 1: Because of the instantaneous nature of runs, the equity holders cannot provide the bank with new equity when a run occurs.

Assumption 2: As soon as the condition for a run to be possible, $S_t \leq \overline{S}_t = X(t)/((1-\alpha)(1+x))$, is satisfied, the run will occur, the bank will have to be liquidated and the equity holders will get nothing.

Assumption 3: If no run occurs and the bank wants to liquidate its projects, it can do so at a proportional variable cost of β, where $\beta < \alpha$ reflects the fact that assets can be sold at a higher price when liquidation occurs voluntarily than in the event of a run.

Under these additional assumptions, the bank with two depositors in effect holds two perpetual down-and-out call options on $(1-\beta)(1+x)S_t$ with an exercise price of $X(t)$ and a knockout price of[1]

$$K(t) = \frac{1-\beta}{1-\alpha} X(t).$$ (9)

Let $C_\infty\big((1-\beta)(1+x)S_t; K(t)\big)$ denote the value of the perpetual down-and-out call option. Making the change in variables

$$V = \frac{(1-\beta)(1+x)S_t}{X(t)}$$ (10)

and defining

$$F(V) = \frac{C_\infty}{X(t)},$$ (11)

F satisfies the following ordinary differential equation:

---

[1] From Assumption 2, a run occurs whenever $S_t = \overline{S}_t = X(t)/((1-\alpha)(1+x))$. Hence, in units of $(1-\beta)(1+x)S_t$, the knock-out price is given by

$$K(t) = (1-\beta)(1+x)\overline{S}_t = (1-\beta)(1+x)\frac{X(t)}{(1-\alpha)(1+x)} = \frac{1-\beta}{1-\alpha}X(t).$$

$$\tfrac{1}{2}\sigma^2 V^2 F'' + (r - r^*)VF' - (r - r^*)F = 0,\qquad(12)$$

subject to the boundary condition:[2]

$$F\left(\frac{1-\beta}{1-\alpha}\right) = 0.\qquad(13)$$

The solution is

$$F(V) = V - \left(\frac{1-\beta}{1-\alpha}\right)^{1+\gamma^*}\cdot V^{-\gamma^*},\quad \gamma^* \equiv 2\frac{r-r^*}{\sigma^2}.\qquad(14)$$

Substituting the original variables back into (14) yields

$$C_\infty = F(V)X(t)$$

$$= VX(t) - \left(\frac{1-\beta}{1-\alpha}\right)^{1+\gamma^*} X(t)V^{-\gamma^*}$$

$$= (1-\beta)(1+x)S_t - \left(\frac{1-\beta}{1-\alpha}\right)^{1+\gamma^*} X(t)\left(\frac{(1-\beta)(1+x)S_t}{X(t)}\right)^{-\gamma^*}\qquad(15)$$

$$= (1-\beta)(1+x)\left[S_t - \left(\frac{X(t)}{(1-\alpha)(1+x)}\right)^{1+\gamma^*} S_t^{-\gamma^*}\right].$$

Equation (15) gives the value of the bank's equity when depositors might choose to run on the bank.[3] It equals asset value net of liquidation costs, $(1-\beta)(1+x)S_t$, minus the expected losses resulting from a run, which are equal to the discount resulting from the knock-out feature of the option. An example of the dependence of equity value on asset value is given in Figures 5.3 and 5.4. It is interesting to note that equity value is increasing in the deposit spread $r-r^*$ and decreasing in asset risk σ.

5.6 The Shareholders' Recapitalization Decision

The above derivation of the value of equity assumed that the shareholders would not recapitalize the bank if asset value falls and a bank run is imminent. This additional assumption was justified with the argument that bank runs occur quickly, and therefore a recapitalization is impossible. It might, however, be interesting to analyze the question of whether shareholders *would be willing* to recapitalize the bank.

---

[2] On the pricing of knock-out options see Ingersoll (1987), p. 371 f.
[3] In the sequel, equity value is assumed to be given by (15), although actual equity value is double that amount. This additional assumption has no influence on the results of the following sections.

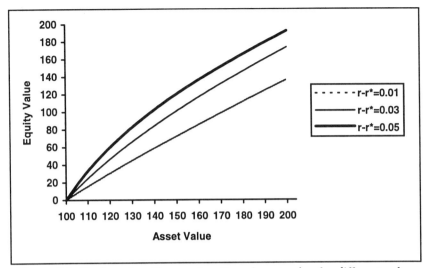

Figure 5.3: *Equity value C_∞ as a function of asset value for different values of the deposit spread $r - r*$ and for the following parameter values: $\alpha = 0.1$, $\beta = 0.05$, $x = 0.\overline{1}$, $X = 100$ and $\sigma = 0.2$. As asset value rises, so does equity value. Moreover, equity value is higher, the wider the deposit spread.*

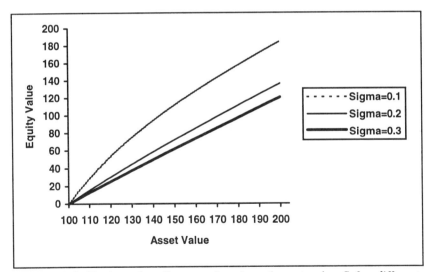

Figure 5.4: *Equity value C_∞ as a function of asset value S for different values of asset risk σ and for the following parameter values: $\alpha = 0.1$, $\beta = 0.05$, $x = 0.\overline{1}$, $X = 100$ and $r - r* = 0.01$. As asset value rises, so does the value of equity. A rising asset risk, however, has a negative influence on equity value.*

To answer it, we model the recapitalization decision as an impulse on S that results in an increase in asset value from $(1+x)S_t$ to $(1+w)(1+x)S_t$, where $w>0$ denotes the percentage increase in asset value achieved through recapitalization.[4] The value of equity after recapitalization is

$$C_\infty = (1-\beta)(1+x)\left[(1+w)S_t - \left(\frac{X(t)}{(1-\alpha)(1+x)}\right)^{1+\gamma^*} \cdot \left((1+w)S_t\right)^{-\gamma^*}\right]. \quad (16)$$

The increase in the value of equity equals

$$\Delta C_\infty = (1-\beta)(1+x)\left[wS_t - \left(\frac{X(t)}{(1-\alpha)(1+x)}\right)^{1+\gamma^*}\left((1+w)^{-\gamma^*}-1\right)S_t^{-\gamma^*}\right]. \quad (17)$$

Increasing asset value is only profitable for shareholders if the increase in equity value exceeds the initial investment $w(1+x)S_t$, that is, if

$$\Delta C_\infty > w(1+x)S_t. \quad (18)$$

Condition (18) can be written as

$$(1-\beta)(1+x)\left[wS_t - \left(\frac{X(t)}{(1-\alpha)(1+x)}\right)^{1+\gamma^*} \cdot \left((1+w)^{-\gamma^*}-1\right)S_t^{-\gamma^*}\right]$$
$$> w(1+x)S_t, \quad (19)$$

or

$$(1-\beta)(1+x)\left(\frac{X(t)}{(1-\alpha)(1+x)}\right)^{1+\gamma^*} \cdot \left(1-(1+w)^{-\gamma^*}\right)S_t^{-\gamma^*} > \beta w(1+x)S_t, \quad (20)$$

which says that the increase in equity value resulting from making a run less likely must exceed the value lost because of liquidation costs, $\beta w(1+x)S_t$. Equivalently, (20) can be written as

$$\left(\frac{S_t}{\frac{X(t)}{(1-\alpha)(1+x)}}\right)^{1+\gamma^*} < \frac{1-\beta}{\beta}\frac{1-(1+w)^{-\gamma^*}}{w}. \quad (21)$$

Equation (21) shows that shareholders will choose to recapitalize the bank as soon as asset value is sufficiently low, as Figure 5.5 demonstrates.
The result that shareholders would be willing to recapitalize the bank is in sharp contrast to that of underinvestment encountered in Chapter 3. To understand why, remember that the underinvestment result of Section

---

[4] Note the similarity of this analysis with that presented in Section 3.5.1 (Myers' underinvestment problem).

3.5.1 obtained because a part of the increase in firm value resulting from the recapitalization accrued to debt holders. Here, however, recapitalizing the bank does not result in a wealth transfer to the depositors.

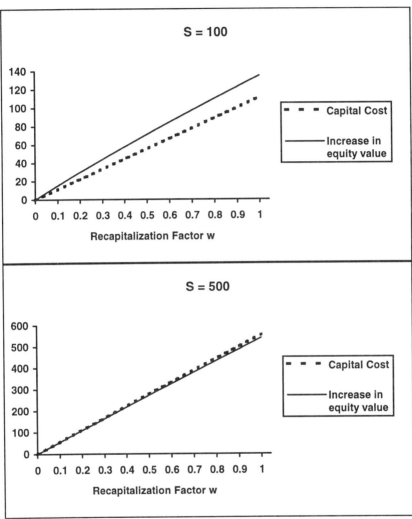

Figure 5.5: *Increase in equity value as a function of the recapitalization share w for different current asset values and for the following parameter values:* $\alpha = 0.1$, $\beta = 0.05$, $x = 0.\overline{1}$, $X = 100$, $r - r^* = 0.01$ *and* $\sigma = 0.2$. *If asset value S is low at 100 (upper panel), shareholders choose to recapitalize the bank. If asset value is high at 500 (lower panel), recapitalization is not profitable, because the capital cost exceeds the resulting increase in equity value.*

5.7 The Bank's Investment Incentives when Bank Runs are Possible

Once bank equity has been valued taking the depositors' run decision into account, the bank's investment incentives can be analyzed. The basic question to be answered in this respect is that of knowing how the possibility of a bank run influences the bank's risk-taking incentives. Again, assume that shareholders cannot recapitalize the bank if asset value falls. Then, bank equity value is given by (15). Taking the partial derivative of this expression with respect to $\gamma *$ yields

$$\frac{\partial C_\infty}{\partial \gamma *} = -(1-\beta)(1+x)\left(\frac{X(t)}{(1-\alpha)(1+x)}\right)^{1+\gamma*} S_t^{-\gamma*} \ln\left(\frac{\dfrac{X(t)}{(1-\alpha)(1+x)}}{S_t}\right), \quad (22)$$

which is positive since (by Assumption 2 above) the bank only exists as long as $S_t > X(t)/((1-\alpha)(1+x))$, so that

$$\ln\left(\frac{\dfrac{X(t)}{(1-\alpha)(1+x)}}{S_t}\right) < 0. \quad (23)$$

What insight can be gained from equation (22)? Remembering that

$$\frac{d\gamma *}{d\sigma^2} = -\frac{\gamma *}{\sigma^2} < 0, \quad (24)$$

we have

$$\frac{\partial C_\infty}{\partial \sigma^2} = \frac{\partial C_\infty}{\partial \gamma *}\frac{d\gamma *}{d\sigma^2} < 0. \quad (25)$$

Equation (25) implies that, as long as assets are fairly priced, the bank is better off by reducing asset risk as much as possible, that is, by setting $\sigma = 0$ in the limit. In other words, *the possibility of bank runs leads the bank to reduce its risk.*

This fact yields an interesting endogenous justification for the existence of demandable debt: demandable debt resolves the incentive problem of risk-shifting by allowing the depositors to withdraw as soon as the value of the bank's investment falls to a prespecified level. This result complements the classical explanation for demandable debt found in the literature, namely that of allowing depositors to react to stochastic preference shocks and liquidity needs by providing them with consumption flexibility. It extends the results of Calomiris and Kahn (1991), who show that demandable debt can be thought of as an incentive scheme to deter bank management from fraud. It demonstrates the disciplining effect of the possibility of bank runs that has been mentioned by several authors in the literature.[5]

The result of equation (25) has, however, another intuitive interpretation. By allowing depositors to withdraw at any time, demandable debt makes the claim of the depositors riskless. Hence, the deposit contract allows a *separation* of the returns for the time value of money and for the riskiness of the underlying venture. The latter is borne entirely by the equity holders.[6]

Because of equation (25), the bank might decide to invest everything in the risk-free asset. For completeness, let us compute the value of equity in this case. By assumption, the risk-free asset can be liquidated at no cost. Let B_0 be the initial amount invested in the risk-free asset and $B(t)$ denote the value of the risk-free asset at time t. Then,

$$B(t) = B_0 e^{rt}. \tag{26}$$

With a capital x, total asset value equals

$$(1+x)B(t) = (1+x)B_0 e^{rt}. \tag{27}$$

Now, the payoff to the equity holders if they choose to liquidate the bank at time t is

$$L(t) = (1+x)B(t) - X(t) = (1+x)B_0 e^{rt} - X_0 e^{r^*t}. \tag{28}$$

By assumption, $B_0 = X_0$, so

$$L(t) = X_0\left((1+x)e^{rt} - e^{r^*t}\right). \tag{29}$$

The present value of this expression is

$$L_0(t) = X_0\left((1+x) - e^{(r^*-r)t}\right). \tag{30}$$

To determine when the equity holders will choose to liquidate the bank, compute

$$\frac{\partial L_0(t)}{\partial t} = -(r^*-r)X_0 e^{(r^*-r)t} = (r-r^*)X_0 e^{(r^*-r)t} > 0. \tag{31}$$

Equation (31) implies that equity holders will never choose to liquidate the bank. Therefore, the value of equity is

$$L = \lim_{t\uparrow\infty} L_0(t) = \lim_{t\uparrow\infty} X_0\left((1+x) - e^{(r^*-r)t}\right) = X_0(1+x). \tag{32}$$

Equation (32) gives the value of equity when the bank invests everything in the risk-free asset. One easily sees that, at initial time, when the investment decision is made,

---

[5] See, for example, Kaufman (1988), p. 568 and Baer and Brewer (1986).
[6] See Postlewaite and Vives (1987), p. 490.

$$L = X_0(1+x) > C_\infty = (1-\beta)(1+x)X_0\left(1-\left(\frac{1}{(1-\alpha)(1+x)}\right)^{1+\gamma^*}\right), \quad (33)$$

so that the bank will decide to invest everything in the risk-free asset. As a consequence, in equilibrium, runs never occur.

5.8 The Bank's Funding Decision

Once the bank's optimal investment choice has been determined, one can work backward through the game and analyze the equity holders' financing decision, i.e. determine what amount of capital x equity holders will choose to provide to the bank in order to maximize their expected profit. Section 5.8.1 discusses the profitability of intermediation in general. Sections 5.8.2 and 5.8.3 then explore the equity holders' optimal funding policies for positive and zero asset risk, respectively.

5.8.1 On the Feasibility of Viable Intermediation in General

For the equity holders to be ready to provide some capital, the value of bank equity at initial time must exceed the capital cost xX_0. In other words, the expected profit from the intermediation activity, $G = C_\infty - xX_0$, must be positive. Treating the general case with positive variance first, and using the fact that $S_0 = X_0$, we get

$$G = C_\infty - xX_0$$

$$= (1-\beta)(1+x)\left(X_0 - \left(\frac{X_0}{(1-\alpha)(1+x)}\right)^{1+\gamma^*} X_0^{-\gamma^*}\right) - xX_0 \quad (34)$$

$$= X_0\left(1 - \beta(1+x) - \frac{(1-\beta)(1+x)}{((1-\alpha)(1+x))^{1+\gamma^*}}\right).$$

For G to be positive, we must have

$$1 - \beta(1+x) - \frac{(1-\beta)(1+x)}{((1-\alpha)(1+x))^{1+\gamma^*}} > 0, \quad (35)$$

that is,

$$\frac{(1-\beta)(1+x)}{1-\beta(1+x)} < ((1-\alpha)(1+x))^{1+\gamma^*}, \quad (36)$$

or

$$1 + \gamma^* > \frac{\ln((1-\beta)(1+x)) - \ln(1-\beta(1+x))}{\ln((1-\alpha)(1+x))}. \quad (37)$$

Equation (37) means that, for the expected profit from funding a bank to be positive, the deposit spread must exceed a certain value, which, for a given capital x, is proportional to the variance of the return on the bank's investments:

$$1+2\frac{r-r*}{\sigma^2} > \frac{\ln\left((1-\beta)(1+x)\right)-\ln\left(1-\beta(1+x)\right)}{\ln\left((1-\alpha)(1+x)\right)} \Leftrightarrow$$

$$r-r* > \frac{\sigma^2}{2}\left(\frac{\ln\left((1-\beta)(1+x)\right)-\ln\left(1-\beta(1+x)\right)}{\ln\left((1-\alpha)(1+x)\right)}-1\right).$$

(38)

From equation (38), it is clear that the bank will be ready to provide some starting capital if the deposit spread is high enough. This result is illustrated in Figure 5.6.

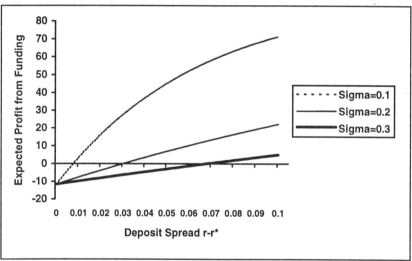

Figure 5.6: *Expected profit from funding a bank as a function of the deposit spread $r - r*$ for different values of asset risk σ and for the following parameter values: $\alpha = 0.1$, $\beta = 0.05$, $x = 0.2$ and $X = 100$. The expected profit is higher, the greater the deposit spread and the lower the asset risk.*

5.8.2 Optimal Bank Capital when Asset Risk is Positive

Now that the feasibility of viable intermediation has been demonstrated, the equity holders' optimal funding policy can be analyzed in more detail. For completeness, let us first explore the case in which asset risk σ is positive. From equation (34) above,

$$\frac{\partial G}{\partial x} = X_0\left(-\beta + \gamma * \cdot \frac{1-\beta}{\left((1-\alpha)(1+x)\right)^{1+\gamma*}}\right).\qquad(39)$$

Setting this expression equal to zero and solving for x yields[7]

$$(1-\alpha)(1+x) = \left(\gamma * \cdot \frac{1-\beta}{\beta}\right)^{\frac{1}{1+\gamma*}} \quad\Leftrightarrow\quad x = \frac{\left(\gamma *(1-\beta)/\beta\right)^{\frac{1}{1+\gamma*}}}{1-\alpha} - 1.\quad(40)$$

Let $\bar{x}$ denote the optimal value of x as given by (40). Then,

$$\frac{\partial \bar{x}}{\partial \alpha} = \frac{1}{(1-\alpha)^2}\left(\gamma * \cdot \frac{1-\beta}{\beta}\right)^{\frac{1}{1+\gamma*}} > 0 \qquad(41)$$

and

$$\frac{\partial \bar{x}}{\partial \gamma *} = \frac{1}{1-\alpha}\left(\gamma * \cdot \frac{1-\beta}{\beta}\right)^{\frac{1}{1+\gamma*}}\frac{1}{1+\gamma*}\left(\frac{1}{\gamma*} - \frac{\ln\left(\gamma * \cdot \frac{1-\beta}{\beta}\right)}{(1+\gamma*)}\right).\qquad(42)$$

Equation (41) says that an increase in α, the proportional liquidation cost in the event of a run, leads the equity holders to provide the bank with *more* capital. The sign of expression (42) is ambiguous. It is, therefore, not possible to say how a change in the deposit spread or asset risk (which both enter $\gamma *$) will, in general, influence the optimal capital x. The reason for this ambiguity is that an increase in the deposit spread or a reduction in asset risk has two conflicting effects. On the one hand, it makes intermediation more profitable, which would call for a higher capital commitment. On the other, it makes runs less probable, thus reducing the capital required at initial time. The question of which of these effects dominates can only be answered on a case-by-case basis.

As shown previously, the bank will seek to set $\gamma *$ as high as possible by reducing σ. Hence, the optimal funding share x can be determined by taking the limit as $\gamma * \uparrow \infty$ of expression (40), yielding

[7] Since

$$\frac{\partial^2 G}{\partial x^2} = -X_0\gamma *(1+\gamma*)\cdot\frac{1-\beta}{(1-\alpha)^{1+\gamma*}(1+x)^{2+\gamma*}} < 0,$$

one is ensured that condition (40) is that required for a maximum.

$$(1-\alpha)(1+x) = \lim_{\gamma^*\uparrow\infty}\left(\gamma^*\cdot\frac{1-\beta}{\beta}\right)^{\frac{1}{1+\gamma^*}} = e^{\lim_{\gamma^*\uparrow\infty}\frac{\ln\left(\gamma^*\cdot\frac{1-\beta}{\beta}\right)}{1+\gamma^*}} = e^{\lim_{\gamma^*\uparrow\infty}\frac{1/\gamma^*}{1}} = e^0 = 1, \quad (43)$$

which means that the profit-maximizing x is set such that the depositors just agree to deposit their money, and everything is invested in an almost risk-free asset.

To confirm the result of Section 5.8.1 that funding a bank can be profitable for equity holders, compute

$$\lim_{\sigma^2\downarrow 0} G = \lim_{\gamma^*\uparrow\infty} X_0\left(1-\beta(1+x) - \frac{(1-\beta)(1+x)}{\left((1-\alpha)(1+x)\right)^{1+\gamma^*}}\right). \quad (44)$$

By assumption, $\left((1-\alpha)(1+x)\right) \geq 1$, so two cases have to be distinguished:

- Case 1: $\left((1-\alpha)(1+x)\right) = 1$

Then,

$$\lim_{\sigma^2\rightarrow 0} G = X_0\left(1-\beta(1+x)-(1-\beta)(1+x)\right) = -xX_0 < 0. \quad (45)$$

The bank cannot make money, and hence the equity owners have no incentive to provide the bank with capital. This situation is in fact the degenerate case where the depositors deposit their money and withdraw it immediately, thus triggering costly liquidation and making intermediation unprofitable.

- Case 2: $\left((1-\alpha)(1+x)\right) > 1$

Then,

$$\lim_{\sigma^2\rightarrow 0} G = X_0\left(1-\beta(1+x)\right). \quad (46)$$

This value will be positive as long as $x < (1-\beta)/\beta$. The question of the feasibility of a bank then boils down to the question of knowing if the above condition is satisfied. Since the condition $x > \alpha/(1-\alpha)$ must be satisfied to induce potential depositors to deposit, the bank can only be viable if the condition $(1-\beta)/\beta > x > \alpha/(1-\alpha)$ is realized. This is only possible if $\alpha + \beta < 1$, that is, if liquidation costs are not too high.

5.8.3 Optimal Bank Capital with Zero Asset Risk

Turning to the case where asset risk σ is zero, the value of the bank's expected profit, $G = L - xX_0$, can be computed using (32) to yield:

$$G = L - xX_0 = (1+x)X_0 - xX_0 = X_0. \quad (46)$$

Clearly,

$$\frac{\partial G}{\partial x} = 0, \tag{47}$$

so that the equity holders are indifferent so as to which amount of capital to invest into the bank. This is not surprising: since everything is invested in the risk-free asset, which earns a return above the deposit rate r^* and can be liquidated at no cost, runs never occur. Therefore, bank capital, which was introduced as a means to cover expected liquidation costs and lead potential depositors to deposit, is no more necessary and becomes irrelevant.

5.9 A Note on the Equilibrium Deposit Spread

The analysis above took the deposit spread as given. How high should one expect it to be? If entry into the banking market is free, then one should expect the deposit spread to be zero. Therefore, for the expected profits to financial intermediation to be positive, entry restrictions, such as chartering, are necessary.

For intermediation to yield positive expected profits, another condition closely related to restricted entry is required. To see this, suppose that the risk-free asset could also be traded at no cost by depositors. Then, they would be able to achieve a return of r on their own, and there would be no reason for banks to exist. It is not unreasonable to assume, however, that some economic agents have a cost advantage in asset trading and therefore become financial intermediaries. In this case, the bank provides liquidity services to depositors, which then agree to receive an interest $r^* < r$ on their deposits. This transactions costs rationale for specialization has been stressed by Merton (1989).

5.10 Conclusion

The analysis of this chapter has demonstrated that a bank will be vulnerable to a bank as soon as the value of its assets net of liquidation costs falls below the face value of deposits. When costs have to be incurred in the event of liquidation of bank assets, the bank has to provide capital covering these expected costs in order to induce depositors to deposit their money. In this sense, banks have a role as providers of *capital*.

Given the depositors' optimal strategy as to when to run on the bank, the bank's equity can be valued as a knock-out perpetual call option. It has been shown that, when asset value falls between a certain level, shareholders would usually be willing to recapitalize the bank.

The bank behaves so as to maximize equity value, taking into account the possibility of runs. When making its investment decision, it *anticipates* the possibility of runs and reduces the risk of its investments as much as possible in order to avoid runs. In other words, the possibility of bank runs disciplines bank behavior. Demandable debt can thus be understood as an optimal contractual arrangement designed to deter banks from engaging in risk-shifting activities. The simple model presented above provides a very strong illustration of this fact: the bank's investment decision has no interior optimum and everything is invested in the risk-free asset. Thus, runs never occur in equilibrium. A somewhat richer, but analytically more demanding model could consider a portfolio of riskless and risky asset and show that the share invested in the risky asset decreases when bank runs are possible.

Turning to the *financing* decision, it was demonstrated that financial intermediation will be profitable as long as the deposit spread is sufficiently high. For such to be the case, barriers to entry, such as chartering or transactions costs advantages, are necessary. When asset risk is positive, optimal bank capital increases with liquidation costs. The dependence of optimal bank capital on the deposit spread and asset risk, however, is ambiguous. The reason for this ambiguity is that an increase in the deposit spread or a reduction in asset risk has two conflicting effects. On the one hand, it makes intermediation more profitable, which would call for a higher capital commitment. On the other, it makes runs less probable, thus reducing the capital required at initial time. When asset risk is zero, shareholders are indifferent as to which amount of bank capital to invest into the bank.

6. Deposit Insurance

6.1 Introduction

The analysis in Chapter 5 considered bank runs and showed that their possibility might lead the bank to invest only in the risk-free asset if there is no underpricing of risky assets. A natural question that arises is that of knowing if the bank can prevent runs by another means than providing capital. This chapter analyzes one of these possibilities: deposit insurance.

Deposit insurance ensures the depositors that they will not suffer if others choose to withdraw their money and they do not. Under the simplifying assumptions of Chapter 5, deposit insurance means that the depositors are sure they will receive at least $X(t) = X_0 e^{r^* t}$ whenever they choose to withdraw. Hence, the incentive to try to withdraw first embedded in condition (6) of Chapter 5 disappears and depositors would, in this model, never choose to withdraw.

In the last few years, a sizeable literature has emerged which is devoted to deposit insurance pricing. While fair pricing, which was developed in the wave of the Savings and Loans debacle, is an economically important issue, the nature of the insurance premium as a sunk cost for the bank has the important consequence that it cannot be expected to change the incentives faced by the bank. This important point was stressed by John et al. (1991), who demonstrated that the risk-shifting incentives of a depository institution fundamentally arise from the existence of limited liability and the associated convex payoff to equity holders. Their analysis constitutes the starting point for the issues addressed in this chapter:

- How does the existence of deposit insurance influence the risk-shifting incentives faced by the bank and its funding decision?
- What is the economic cost or benefit of deposit insurance?

6.2 The Model

Consider a bank operating in conditions similar to those of Chapter 5: At time t, it has a total amount of $X(t) = X_0 e^{r^* t}$ in deposits. Asset value S follows a geometric Brownian motion. For each dollar deposited at initial time 0, bank equity holders have added an amount x in equity capital. Hence, total asset value is given by $(1+x)S_t$. To liquidate the assets, a proportional cost of β has to be incurred. In the event of a bank run, this proportional liquidation cost is higher and amounts to α.

Suppose that deposits are insured. At any point in time, the guarantor knows the value of the bank's assets and can liquidate them immediately if he wishes. Assume, further, that if the guarantor decides to seize the assets and to liquidate them, he will realize $(1-\beta)(1+x)S_t$. If liquidation occurs, depositors are paid in full; they receive the face amount of their deposits, $X(t)$, regardless of the proceeds from liquidation. Any proceeds from liquidation in excess of the face value of deposits $X(t)$ are paid to the bank's shareholders. Thus, the guarantor in effect writes a perpetual put option on the bank's assets, with an exercise price equal to the current value of deposits, $X(t)$.

Under these assumptions, the structure of the game is that given in Figure 6.1. After the bank is funded, it makes its investment decision. The guarantor observes asset value and may choose to liquidate the bank.

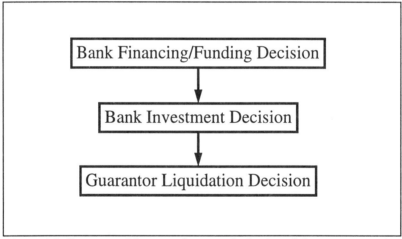

Figure 6.1: Structure of the game between the bank and the guarantor when the latter can observe asset value perfectly and liquidate assets immediately. After the bank is funded, it chooses its investment strategy. If asset value reaches a certain level, the guarantor may decide to liquidate the bank.

The structure of the chapter is as follows: Section 6.3 uses option pricing to value the cost of deposit insurance, bank equity and social welfare. Section 6.4 explores the guarantor's optimal liquidation strategy and comments on the ability of deposit insurance to enhance social welfare. Section 6.5 addresses the question of whether the existence of deposit insurance influences the bank's investment and financing decisions. Sections 6.6 and 6.7 extend the basic model to the cases where immediate liquidation is not possible and where the guarantor cannot observe asset value, respectively. Section 6.8 concludes the chapter.

6.3 Valuing Deposit Insurance, Bank Equity and Social Welfare

Before the game depicted in Figure 6.1 can be solved, the players' payoffs must be valued using option pricing. This is done in Sections 6.3.1 and 6.3.2. In Section 6.3.3, a measure of social welfare is presented.

6.3.1 The Value of the Deposit Insurance Guarantee

Remember that the guarantee has the structure of a perpetual put option on $(1-\beta)(1+x)S_t$ with an exercise price equal to the face value of deposits, $X(t)$. Let $P_\infty((1-\beta)(1+x)S_t,\overline{S}(X(t)))$ denote the value of the put option, where $\overline{S}(X(t))$ denotes the asset value at which the guarantor liquidates the bank.

Making the change in variables

$$V = \frac{(1-\beta)(1+x)S_t}{X(t)} \tag{1}$$

and defining

$$F(V) = \frac{P_\infty}{X(t)}, \tag{2}$$

F satisfies the following ordinary differential equation:[1]

$$\tfrac{1}{2}\sigma^2 V^2 F'' + (r-r^*)VF' - (r-r^*)F = 0, \tag{3}$$

subject to the boundary conditions

$$F(\infty) = 0, \tag{4}$$

$$F(\overline{V}) = 1 - \overline{V} = 1 - \frac{(1-\beta)(1+x)\overline{S}(X(t))}{X(t)}, \tag{5}$$

where $\overline{V}$ denotes the assets-to-deposits ratio at which the bank is closed by the guarantor.[2] Notice that liquidation costs, β, and bank capital, x, have been taken care of in the definition of $\overline{V}$. The general solution to (3) is

$$F(V) = \alpha_1 V + \alpha_2 V^{-\gamma^*}, \quad \gamma^* \equiv 2\frac{r-r^*}{\sigma^2}. \tag{6}$$

---

[1] Merton (1990), p. 298, uses a similar procedure to price a warrant with a continuously changing exercise price.

[2] As will be seen shortly, $\overline{S}(X(t))$ is linear in $X(t)$, so that $\overline{V}$ does not depend on $X(t)$.

From boundary condition (4), we can easily see that $\alpha_1 = 0$. From boundary condition (5), we get

$$F(\overline{V}) = 1 - \overline{V} = \alpha_2 \overline{V}^{-\gamma^*},$$ (7)

and therefore:

$$\alpha_2 = \left(1 - \overline{V}\right) \cdot \overline{V}^{\gamma^*}.$$ (8)

Hence,

$$F = \alpha_2 \cdot V = \left(1 - \overline{V}\right) \cdot \left(\frac{V}{\overline{V}}\right)^{-\gamma^*}.$$ (9)

From the definition of F, we get

$$P_\infty = F(V) \cdot X(t)$$

$$= X(t)\left(1 - \overline{V}\right) \cdot \left(\frac{V}{\overline{V}}\right)^{-\gamma^*}$$ (10)

$$= \left(X(t) - (1 - \beta)(1 + x)\overline{S}(X(t))\right) \cdot \left(\frac{S_t}{\overline{S}(X(t))}\right)^{-\gamma^*}.$$

Equation (10) gives the expected cost of the deposit insurance guarantee as a function of the guarantor's liquidation strategy, $\overline{S}(X(t))$. This cost equals the amount paid out by the insurer in the event of liquidation, $X(t) - (1 - \beta)(1 + x)\overline{S}(X(t))$, times the factor $(S_t / \overline{S}(X(t)))^{-\gamma^*}$, which takes both the risk-neutral probability of liquidation and the time value of money into account.

6.3.2 The Value of Bank Equity

Bank equity can be modeled as a perpetual knock-out call option on $(1 - \beta)(1 + x)S_t$ with a knock-out price equal to the guarantors' liquidation strategy $\overline{S}(X(t))$. Let $C_\infty\left((1 - \beta)(1 + x)S_t; \overline{S}(X(t))\right)$ denote the value of this perpetual down-and-out option. Making the change in variables (1) and defining

$$F(V) = \frac{C_\infty}{X(t)},$$ (11)

F satisfies (3) subject to the boundary condition

$$F(\overline{V}) = 0.$$ (12)

Using (6) and setting $\alpha_1 = 1$ to avoid arbitrage, boundary condition (12) yields

$$F(\overline{V}) = 0 = \overline{V} + \alpha_2 \overline{V}^{-\gamma^*}, \tag{13}$$

and therefore:

$$\alpha_2 = -\overline{V}^{1+\gamma^*}. \tag{14}$$

Hence,

$$F = V + \alpha_2 \cdot V^{-\gamma^*} = V - \overline{V} \cdot \left(\frac{V}{\overline{V}}\right)^{-\gamma^*}. \tag{15}$$

From the definition of F, we get

$$
\begin{aligned}
C_\infty &= F(V) \cdot X(t) \\
&= X(t)\left(V - \overline{V}\left(\frac{V}{\overline{V}}\right)^{-\gamma^*}\right) \\
&= (1-\beta)(1+x)\left(S_t - \overline{S}(X(t))\left(\frac{S_t}{\overline{S}(X(t))}\right)^{-\gamma^*}\right).
\end{aligned}
\tag{16}
$$

Equation (16) gives the value of bank equity as a function of the guarantor's liquidation strategy, $\overline{S}(X(t))$. This value equals total asset value net of liquidation costs, $(1-\beta)(1+x)S$, minus the discount resulting from the knock-out feature of the option, $(1-\beta)(1+x)\overline{S}(X(t)) \cdot (S_t / \overline{S}(X(t)))^{-\gamma^*}$. Note, again, that $(S_t / \overline{S}(X(t)))^{-\gamma^*}$ can be thought of as the risk-neutral probability of liquidation, adjusted for the time value of money.

6.3.3 The Value of Social Welfare

To analyze the welfare effects of deposit insurance, social welfare must be valued. An appealing measure of social welfare is the difference between the bank's equity value, C_∞, and the value of the deposit insurance guarantee, P_∞. Using (10) and (16), the value of the social surplus, Π, is given by:

$$
\begin{aligned}
\Pi &= C_\infty - P_\infty \\
&= X(t)\left(V - \overline{V}\left(\frac{V}{\overline{V}}\right)^{-\gamma^*}\right) - X(t)\left(1 - \overline{V}\right)\left(\frac{V}{\overline{V}}\right)^{-\gamma^*} \\
&= X(t)\left(V - \left(\frac{V}{\overline{V}}\right)^{-\gamma^*}\right) = (1-\beta)(1+x)S_t - X(t)\left(\frac{S}{\overline{S}}\right)^{-\gamma^*}.
\end{aligned}
\tag{17}
$$

The value of social surplus as given by (17) equals asset value $(1-\beta)(1+x)S$ minus the expected payout to depositors, $X(t)(S_t/\overline{S})^{-\gamma^*}$. As usual, the factor $(S_t/\overline{S})^{-\gamma^*}$ can be interpreted as the risk-neutral probability of liquidation, adjusted for the time value of money.

Note that expressions (10), (16) and (17) are much simpler when they are represented using V. Consequently, this form shall sometimes be used in the sequel.

6.4 The Guarantor's Liquidation Strategy and Social Welfare

In this section, the guarantor's liquidation strategy and its influence on the value of bank equity and social welfare are discussed. Moreover, the question of whether deposit insurance can enhance social welfare is addressed.

Through his liquidation strategy, the guarantor might follow several objectives. First, he might wish to minimize the *cost* of the deposit insurance guarantee. This question is addressed in section 6.4.1. Second, he might wish to maximize *social welfare*. This problem is discussed in section 6.4.2. Section 6.4.3 then comments on the generic ability of deposit insurance to enhance social welfare.

6.4.1 Minimizing the Cost of the Guarantee

To solve for the optimal exercise strategy from the guarantor's standpoint, one must remember that the guarantor holds a *short* put on the assets. Hence, he will seek to *minimize* its value. This can be achieved by setting

$$\overline{V}=1, \tag{18}$$

which yields a put option value of zero.[3] Equivalently, the cost-minimizing liquidation strategy (18) can be written as:

$$\overline{S}(X(t)) = \frac{X(t)}{(1-\beta)(1+x)}. \tag{19}$$

The intuition for this result is the following: By liquidating bank assets early, the guarantor avoids incurring any liability from the guarantee. The guarantee is, therefore, worthless. As we will see later, however, this liquidation strategy doesn't mean that deposit insurance as a whole is worthless.

---

[3] Note that by setting $\overline{S}=0$, the guarantor also minimizes the value of the guarantee. This result stems from the fact that depositors never withdraw if the bank is not liquidated. A more complicated model with depositors withdrawing their money randomly over time would eliminate this solution.

6.4.2 Maximizing Social Welfare

Suppose now that the guarantor is not concerned with the cost of the guarantee, but chooses a liquidation strategy $\bar{V}$ that maximizes social welfare. Differentiating (17) partially with respect to $\bar{V}$ yields:

$$\frac{\partial \Pi}{\partial \bar{V}} = -\frac{\gamma * X(t)}{\bar{V}}\left(\frac{V}{\bar{V}}\right)^{-\gamma*} < 0 \quad \forall \bar{V} > 0 . \tag{20}$$

Hence, from a social welfare standpoint, it is optimal never to liquidate the bank, that is, to set $\bar{V} = 0$. Note that this in turn implies $\bar{S} = 0$. This result is intuitively clear: by avoiding liquidation altogether, society can save on liquidation costs (which are a loss in this model). Figure 6.2 plots the values of equity C_∞, the guarantee P_∞ and social surplus $\Pi = C_\infty - P_\infty$ as a function of the guarantor's liquidation strategy $\bar{S}$. It illustrates that social surplus is a monotone decreasing function of the liquidation-triggering asset value $\bar{S}$, implying that the socially optimal liquidation strategy for the guarantor is $\bar{S} = 0$.

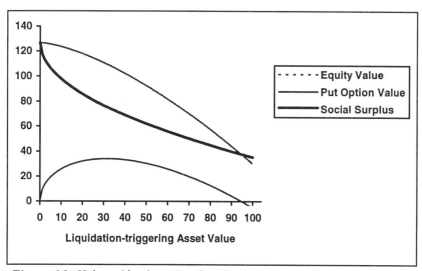

Figure 6.2: *Value of bank equity C_∞, the deposit insurance guarantee P_∞ and social surplus Π as a function of the liquidation-triggering asset value $\bar{S}$ for the following parameter values: $\beta = 0.05$, $x = 0.\bar{1}$, $S = 120$, $X = 100$, $r - r* = 0.01$ and $\sigma = 0.2$. As the liquidation-triggering asset value $\bar{S}$ is increased, both the value of equity and the value of the social surplus fall. The cost of the deposit insurance guarantee rises at first, and then falls. Hence, the socially optimal liquidation strategy for the guarantor is never to liquidate the bank, i.e. set $\bar{S} = 0$.*

6.4.3 Can Deposit Insurance Enhance Social Welfare?

This section is devoted to a very controversial question: can deposit insurance enhance social welfare? And if so, under what conditions? The method followed here is relatively simple. It consists in comparing social welfare (17) with the value of bank equity when there is no deposit insurance.

From the analysis in Chapter 5, remember that when there is no deposit insurance, the value of bank equity equals

$$C_\infty^- = (1-\beta)(1+x)\left(S_t - \left(\frac{X(t)}{(1-\alpha)(1+x)} \right)^{1+\gamma^*} S_t^{-\gamma^*} \right). \qquad (21)$$

For there to be a welfare gain from deposit insurance, social welfare in the presence of deposit insurance must exceed the value of bank equity in the absence of deposit insurance. Formally, one must have

$$\Pi = (1-\beta)(1+x)S_t - X(t)\left(\frac{S_t}{\overline{S}(X(t))} \right)^{-\gamma^*}$$

$$> (1-\beta)(1+x)\left(S_t - \left(\frac{X(t)}{(1-\alpha)(1+x)} \right)^{1+\gamma^*} S_t^{-\gamma^*} \right) = C_\infty^-, \qquad (22)$$

that is

$$X(t)\left(\frac{S_t}{\overline{S}(X(t))} \right)^{-\gamma^*} < (1-\beta)(1+x)\left(\frac{X(t)}{(1-\alpha)(1+x)} \right)^{1+\gamma^*} S_t^{-\gamma^*}, \qquad (23)$$

or

$$\left(\overline{S}(X(t)) \right)^{\gamma^*} < (1-\beta)(1+x)X(t)^{\gamma^*}\left((1-\alpha)(1+x) \right)^{-(1+\gamma^*)}$$

$$= \frac{1-\beta}{1-\alpha}\left(\frac{X(t)}{(1-\alpha)(1+x)} \right)^{\gamma^*}. \qquad (24)$$

In words, deposit insurance will enhance social welfare if the guarantor chooses to liquidate the assets at a *low enough* value. It is interesting to note that the critical liquidation strategy $\overline{S}(X(t))$ under which deposit insurance enhances social welfare depends on two factors: $(1-\beta)/(1-\alpha)$, which measures the value that can be saved by liquidating bank assets normally instead of through a fire sale, and $X(t)/((1-\alpha)(1+x))$, which is the asset value that would trigger a run if deposit insurance did not exist.

To sum up, deposit insurance will create social value either if costly fire sales can be avoided ($\beta < \alpha$), or if the guarantor displays some forbearance (and liquidates less frequently than would be the case if bank runs were possible). This implies that deposit insurance might enhance social welfare even if the guarantor provides no effective guarantee. To see this, suppose the guarantor, seeking to minimize the value of the guarantee, were to set $\overline{S}(X(t)) = X(t)/((1-\beta)(1+x))$, as posited by (19). Substituting this expression into (24) yields the following condition for social welfare enhancement through deposit insurance:

$$\left(\frac{X(t)}{(1-\beta)(1+x)}\right)^{\gamma^*} < \frac{1-\beta}{1-\alpha}\left(\frac{X(t)}{(1-\alpha)(1+x)}\right)^{\gamma^*}. \tag{25}$$

Simplifying (25) then leads to

$$\beta < \alpha, \tag{26}$$

which is satisfied by assumption.

6.5 The Incentive Effects of Deposit Insurance

In this section, the effects of the existence of deposit insurance on the bank's optimal investment and funding decisions are analyzed. Section 6.5.1 shows that the existence of deposit insurance does not alleviate the bank's risk-taking incentives. Section 6.5.2 then demonstrates that deposit insurance and forbearance on the part of the guarantor can be thought of as substitutes for bank capital.

6.5.1 The Investment Decision

A question of interest is that of knowing how the guarantor's liquidation strategy $\overline{V}$ influences the bank's risk-taking incentives. Remembering that

$$\frac{\partial C_\infty}{\partial \sigma^2} = \frac{\partial C_\infty}{\partial \gamma^*}\frac{d\gamma^*}{d\sigma^2} \tag{27}$$

and using

$$\frac{d\gamma^*}{d\sigma^2} = -\frac{\gamma^*}{\sigma^2} < 0 \tag{28}$$

and

$$\frac{\partial C_\infty}{\partial \gamma^*} = X(t)\frac{\partial}{\partial \gamma^*}\left(V - \overline{V}\left(\frac{V}{\overline{V}}\right)^{-\gamma^*}\right)$$

$$= X(t)\overline{V}\left(\frac{V}{\overline{V}}\right)^{-\gamma^*}\ln\left(\frac{V}{\overline{V}}\right) > 0 \quad \forall \overline{V} > 0 \tag{29}$$

yields

$$\frac{\partial C_\infty}{\partial \sigma^2} < 0 \quad \forall \bar{V} > 0. \tag{30}$$

Hence, as long as the assets-to-deposits ratio at which the guarantor chooses to liquidate the bank is positive, the existence of deposit insurance does not alleviate the bank's incentive to reduce its risk. Figure 6.3 plots the value of bank equity C_∞ as a function of the liquidation-triggering assets-to-deposits ratio $\bar{V}$ for different levels of asset risk σ and illustrates that a lowering of $\bar{V}$ raises equity value but does not alter the negative dependence of C_∞ on asset risk σ.

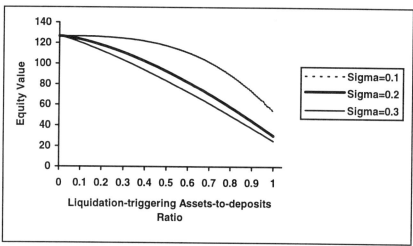

Figure 6.3: *Value of bank equity* C_∞ *as a function of the liquidation-triggering assets-to-deposits ratio* $\bar{V}$ *for different values of asset risk* σ *and the following parameter values:* $\beta = 0.05$, $x = 0.\overline{1}$, $S = 120$, $X = 100$ *and* $r - r^* = 0.01$. *As* $\bar{V}$ *is increased, equity value falls. For any given* $\bar{V}$, *equity value is higher, the lower the asset risk.*

This result is in sharp contrast to the conclusions of the existing literature, which stress the risk-shifting incentives of deposit insurance. Where does the difference come from?

Most of the existing literature implicitly or explicitly assumes that the guarantor cannot monitor the bank perfectly, and therefore cannot assess asset value properly. This alternative assumption is a possible explanation, as will be seen in Section 6.7.

Another important point to be kept in mind when interpreting the results is the assumption implicitly used here that the guarantor chose a liquidation strategy $\bar{V}$, and then stuck to it. There is, however, nothing preventing the guarantor in lowering $\overline{V}$ more and more as the asset value moves closer to the boundary. In the limit, the guarantor sets $\overline{V}$ equal to zero. This lowering of $\overline{V}$ may occur for a number of reasons, the most important being that the guarantor himself usually is an *agent* acting on behalf of the government. Moreover, politicians might be interested in hiding the extent of losses to their voters, which is an additional reason for lowering $\overline{V}$ or avoiding liquidation altogether.[4] Note that, with $\overline{V} = 0$, $\partial C_\infty / \partial \gamma^* = 0$, so that the bank has no incentive to increase or decrease risk.

In the absence of deposit insurance, depositors bear the full cost of losses and therefore have an incentive to withdraw their funds early. This incentive makes the threat of liquidation credible and self-enforcing, thereby leading the bank to reduce its risk. This fact may be an additional reason why the government often fails to liquidate banks early. If the guarantor were to liquidate, banks would tend to invest only in short-term, low-risk assets. This would have important consequences for the real economy, where long-term capital is required. Therefore, without deposit insurance, the possibility might exist that long-term capital becomes unavailable for investment. If the government recognizes this fact, it might wish to make credible that it will not liquidate insolvent banks, thus enabling long-term investment. This point is similar to that made by Diamond and Dybvig (1983).

6.5.2 The Financing Decision

In this section, the influence of the guarantor's liquidation strategy on the optimal amount of capital is analyzed. At the time the financing decision is made, bank equity holders choose a capital x that maximizes their expected profit from funding the bank. Using (16) and setting $S_0 = X_0$ yields:

$$G = C_\infty - xX_0$$

$$= X_0 \left(1 - \beta(1+x) - (1-\beta)(1+x) \left(\frac{\overline{V}}{(1-\beta)(1+x)} \right)^{1+\gamma^*} \right). \tag{31}$$

Differentiating this expression with respect to x and setting the result equal to zero yields

---

[4] See Kane (1995) for an excellent overview of the agency problems of government deposit insurance.

$$(1 - \beta)(1 + x) = \overline{V} \left(\gamma * \frac{1 - \beta}{\beta} \right)^{\frac{1}{1 + \gamma *}}. \tag{32}$$

Hence, x is decreasing in $1 - \overline{V}$, which is a measure of the guarantor's forbearance. The effect of $\overline{V}$ on the optimal capital x and capital share $x / (1 + x)$ is depicted in Figure 6.4.

The analysis in this section demonstrates that deposit insurance and the guarantor's forbearance not only are means of avoiding costly bank runs, fire sales and liquidation. Remember that, in equilibrium, bank runs would almost never occur. Rather, deposit insurance and forbearance might be thought of as *substitutes* for bank capital. It is, therefore, not surprising that deposit insurance was introduced in the Great Depression, in a period of relative capital scarcity. Deposit insurance and forbearance can be thought of as a means of keeping banks alive without requiring huge amounts of capital.

To support this thesis, remember that, before deposit insurance was introduced, banks had much higher capital ratios than today. Kaufman (1988) reports values close to 25 percent at the turn of the last century.

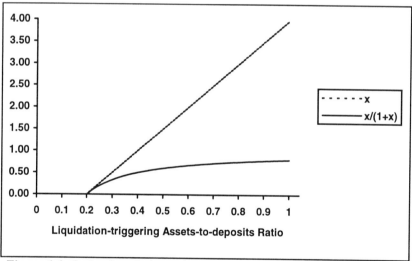

Figure 6.4: *Optimal capital x and capital share $x / (1 + x)$ as a function of the liquidation-triggering assets-to-deposits ratio $\overline{V}$, where x has been constrained to be positive, for the following parameter values: $\beta = 0.05$, $r - r* = 0.01$ and $\sigma = 0.2$. As $\overline{V}$ is increased, optimal capital x and capital share $x / (1 + x)$ rise. Thus, deposit insurance can be considered as a substitute for bank capital.*

6.6 Deposit Insurance when there are Liquidation Delays

The results derived in the previous section depend crucially on the assumptions that asset value is perfectly observable by the guarantor and that liquidation, once decided, can occur immediately. In this section, this latter assumption is relaxed. It is assumed that the guarantor can still observe asset value at no cost, but must wait some time τ before he can liquidate the assets, once he has decided to do so. The case in which the guarantor cannot observe asset value will be discussed in Section 6.7.

In this setting, the game between the guarantor and the bank can be described as follows: as long as the asset value lies above the trigger $\overline{S}(X(t))$, the bank operates freely. If $S = \overline{S}(X(t))$, however, the guarantor decides to liquidate the bank and announces this to the bank. Denote the time at which this occurs by t_0. At this point, the bank may choose to change its investment policy. After the time τ elapses, assets are effectively seized and the bank owners get any amount in excess of the deposit value $X(t_0 + \tau)$. The structure of the game is therefore that depicted in Figure 6.5.

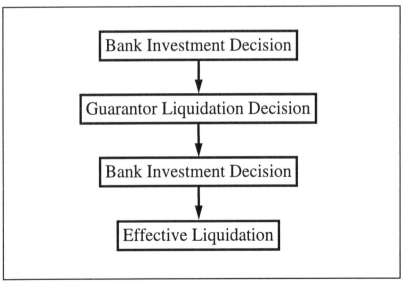

Figure 6.5: Structure of the game when there is a delay of τ between the liquidation decision and effective liquidation. After the bank chooses its investment strategy, the guarantor observes asset value and might choose to liquidate the bank. Then, the bank may choose to change its investment strategy. Finally, after the delay elapses, effective liquidation takes place and payoffs are received.

From the above assumptions, the payoff to the bank's shareholders is

$$Max[0;(1-\beta)(1+x)S_{t_0+\tau} - X(t_0+\tau)], \qquad (33)$$

which has the same structure as the payoff on a call option. The payoff to the guarantor is

$$Min[0;(1-\beta)(1+x)S_{t_0+\tau} - X(t_0+\tau)] \qquad (34)$$

and has the same structure as that of a put option.

To understand the incentive problem resulting from the impossibility of immediate liquidation, consider the risk-taking incentives of the bank. As long as the guarantor has not announced liquidation, the bank has an incentive to *reduce* its risk in order to reduce the probability of the asset value reaching the liquidation trigger. Once liquidation has been decided, however, the bank holds a call option with *finite* life τ and has an incentive to *increase* asset risk to raise the value of the option.

This fact has an important implication for the practice of deposit insurance: to avoid risk-shifting by the bank, the guarantor should monitor the asset value closely and *seize* the assets as soon as he decides to liquidate the bank. This result can be extended to other firms. Many bankruptcy laws, for example, place the firm under judicial administration to avoid the equity holders or managers' engaging in asset substitution or fraud. Our analysis therefore has a very intuitive interpretation.

6.7 Deposit Insurance with Unobservable Asset Value

Suppose now that the guarantor cannot observe the asset value at all and wishes to lead the bank to liquidate voluntarily as soon as the assets-to-deposits ratio V reaches a pre-specified level. Then, he must try to construct an *incentive contract* such that liquidating the bank early is a dominant strategy for the equity holders. The structure of the game, which is depicted in Figure 6.6, stresses that the decision of whether to liquidate the bank now lies in the hands of the equity holders.

6.7.1 A First Attempt: Extending the Model of Chapter 3

One is tempted to use a contract of the type analyzed in Chapter 3, which was shown to be able to lead equity holders to declare bankruptcy voluntarily at any pre-specified level of asset value. In the case of deposit insurance, one would require the bank's equity holders to pay an instantaneous premium of $\phi X(t)dt$ to the insurer, which would have the same revelation function as the interest payment $\phi D(t)dt$ of Chapter 3.

The problem with this approach, however, is that the bank's equity holders might decide to have the *bank* pay the premium with its own assets instead of paying it themselves. This might result in bankruptcy being declared later than the guarantor would expect.

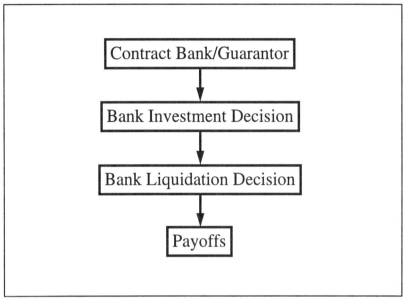

Figure 6.6: *Structure of the game when the guarantor cannot observe asset value. After agreeing on a deposit insurance contract with the guarantor, the bank chooses an investment strategy. At some point, depending on the incentives set by the contractual provisions, the bank might decide to turn its assets to the guarantor for liquidation.*

To see this, suppose that the premium $\phi X(t)dt$ is paid through the sale of bank assets. Then, the value of the bank's assets, S, evolves according to

$$dS = (\mu S - \phi X(t))dt + \sigma S d\tilde{z} , \qquad (35)$$

and the value of bank equity, E, satisfies the following differential equation:

$$\tfrac{1}{2}\sigma^2 S^2 E_{SS} + (rS - \phi X(t))E_S + r^* X(t)E_{X(t)} - rE = 0, \qquad (36)$$

where subscripts to E denote partial derivatives, subject to the following boundary conditions:

$$E(\overline{S}) = 0, \qquad (37)$$

$$E(\infty) = S - \frac{\phi X(t)}{r - r^*}. \qquad (38)$$

Boundary condition (37) states that equity holders get nothing if the bank is closed. Boundary condition (38) states that, as asset value becomes very large, bank closure becomes unlikely and equity value therefore equals asset value minus the present value of insurance premium payments. Making the change in variables

$$V = \frac{S}{X(t)} \tag{39}$$

and defining

$$F(V) = \frac{E(S)}{X(t)}, \tag{40}$$

we get[5]

$$\tfrac{1}{2}\sigma^2 V^2 F'' + ((r - r^*)V - \phi)F' - (r - r^*)F = 0. \tag{41}$$

To solve this equation, we use the same methodology as in Black and Cox (1976) and Hua and Sundaresan (1997). Define

$$Z = \frac{2\phi}{\sigma^2 V} \tag{42}$$

and

$$F(V) = Z^{\gamma^*} e^{-Z} h(Z). \tag{43}$$

Then, (41) reduces to Kummer's equation

$$Zh'' + (2 + \gamma^* - Z)h' - 2h = 0, \tag{44}$$

which has general solution of the form

$$h(Z) = \alpha_1 M(2, 2 + \gamma^*, Z) + \alpha_2 Z^{-(1+\gamma^*)} M(1 - \gamma^*, -\gamma^*, Z), \tag{45}$$

where $M(\cdot, \cdot, \cdot)$ is the confluent hypergeometric function.[6] Substituting the original variables back yields

---

[5] From the definitions in equations (39) and (40), $E_S = F'$, $E_{SS} = F'' / X$, $E_X = F - VF'$. Substituting these expressions into (36) yields

$$\tfrac{1}{2}\sigma^2 S^2 F'' / X + (rS - \phi X)F' + r^* X(F - VF') - rXF = 0.$$

Then, using the definition of V, one obtains

$$X\left(\tfrac{1}{2}\sigma^2 V^2 F'' + (rV - \phi)F' + r^*(F - VF') - rF\right) = 0.$$

Collecting terms and crossing out X then gives (41).

[6] The confluent hypergeometric function $M(\cdot, \cdot, \cdot)$ is defined as

$$M(a, b, x) = 1 + \sum_{n=1}^{\infty} \frac{a_n}{b_n} \frac{x^n}{n!}, \text{ where } a_n = a(a+1)(a+2)...(a+n-1) \text{ and analogously}$$

for b_n. It has the following properties, which are described in Slater (1968):

$$M(a, b, 0) = 1,$$
$$M(a, b, x) = e^x M(b - a, b, -x),$$
$$x^{1-b} M(1 + a - b, 2 - b, x) = x^{1-b} e^x M(1 - a, 2 - b, -x).$$

$$F(V) = Z^{\gamma^*} e^{-Z} h(Z)$$

$$= \alpha_1 \left(\frac{2\phi}{\sigma^2 V} \right)^{\gamma^*} e^{-\frac{2\phi}{\sigma^2 V}} M\left(2,2+\gamma^*, \frac{2\phi}{\sigma^2 V} \right) \tag{46}$$

$$+ \alpha_2 e^{-\frac{2\phi}{\sigma^2 V}} \left(\frac{2\phi}{\sigma^2 V} \right)^{-1} M\left(1-\gamma^*, -\gamma^*, \frac{2\phi}{\sigma^2 V} \right).$$

Using the property $M(a,b,x) = e^x M(b-a,b,-x)$ given in Footnote 6 and remembering that $M(-1,-\gamma^*,x) = 1+x/\gamma^*$, we get

$$F(V) = \alpha_1 \left(\frac{2\phi}{\sigma^2 V} \right)^{\gamma^*} M\left(\gamma^*, 2+\gamma^*, -\frac{2\phi}{\sigma^2 V} \right)$$

$$+ \alpha_2 \frac{\sigma^2 V}{2\phi} M\left(-1, -\gamma^*, -\frac{2\phi}{\sigma^2 V} \right) \tag{47}$$

$$= \alpha_1 \left(\frac{2\phi}{\sigma^2 V} \right)^{\gamma^*} M\left(\gamma^*, 2+\gamma^*, -\frac{2\phi}{\sigma^2 V} \right) + \alpha_2 \frac{\sigma^2 V}{2\phi} \left(1 - \frac{1}{\gamma^*} \frac{2\phi}{\sigma^2 V} \right).$$

Hence,

$$E(S) = F(V) X(t)$$

$$= \alpha_1 X(t) \left(\frac{2\phi X(t)}{\sigma^2 S} \right)^{\gamma^*} M\left(\gamma^*, 2+\gamma^*, -\frac{2\phi X(t)}{\sigma^2 S} \right) \tag{48}$$

$$+ \alpha_2 \frac{\sigma^2 S}{2\phi} \left(1 - \frac{1}{\gamma^*} \frac{2\phi X(t)}{\sigma^2 S} \right).$$

Applying boundary condition (38) yields $\alpha_2 = 2\phi / \sigma^2$. Substituting this result into (48), one obtains

$$E(S) = \alpha_1 X(t) \left(\frac{2\phi X(t)}{\sigma^2 S} \right)^{\gamma^*} M\left(\gamma^*, 2+\gamma^*, -\frac{2\phi X(t)}{\sigma^2 S} \right) + \left(S - \frac{\phi X(t)}{r-r^*} \right). \tag{49}$$

Applying boundary condition (37) then yields

$$\alpha_1 = -\left(\frac{\overline{S}}{X(t)} - \frac{\phi}{r-r^*} \right) \left(\frac{2\phi X(t)}{\sigma^2 \overline{S}} \right)^{-\gamma^*} \Big/ M\left(\gamma^*, 2+\gamma^*, -\frac{2\phi X(t)}{\sigma^2 \overline{S}} \right). \tag{50}$$

Therefore, the value of equity, E, is given by:

$$E(S) = S - \frac{\phi X(t)}{r-r^*} + \left(\frac{\phi X(t)}{r-r^*} - \overline{S} \right) \left(\frac{S}{\overline{S}} \right)^{-\gamma^*} \frac{M\left(\gamma^*, 2+\gamma^*, -\frac{2\phi X(t)}{\sigma^2 S} \right)}{M\left(\gamma^*, 2+\gamma^*, -\frac{2\phi X(t)}{\sigma^2 \overline{S}} \right)}. \tag{51}$$

It is interesting to note that equity value (51) equals asset value S minus $\phi X(t)/(r-r^*)$, the present value of insurance premium payments if bankruptcy never occurs, plus a term that takes the wealth effects of bankruptcy into account.

At this point, the equity holders' *bankruptcy strategy* is still unknown. To determine it, one first removes terms where $\bar{S}$ does not appear in (51), thus obtaining

$$\frac{\bar{S}^{\gamma^*}}{M\left(\gamma^*,2+\gamma^*,-\dfrac{2\phi}{\sigma^2\bar{S}/X(t)}\right)}\left(\frac{\phi X(t)}{r-r^*}-\bar{S}\right). \qquad (52)$$

Taking the logarithmic derivative of (52) with respect to $\bar{S}$ yields the first-order condition

$$\frac{\gamma^*}{\bar{S}}-\frac{M'\left(\gamma^*,2+\gamma^*,-\dfrac{2\phi X(t)}{\sigma^2\bar{S}}\right)\dfrac{2\phi X(t)}{\sigma^2\bar{S}^2}}{M\left(\gamma^*,2+\gamma^*,-\dfrac{2\phi X(t)}{\sigma^2\bar{S}}\right)}+\frac{1}{\bar{S}-\dfrac{\phi X(t)}{r-r^*}}=0, \qquad (53)$$

which can also be written as

$$\left(\frac{\gamma^*}{\bar{S}}+\frac{r-r^*}{(r-r^*)\bar{S}-\phi X(t)}\right)\frac{\sigma^2\bar{S}^2}{2\phi X(t)}=\frac{M'(\cdot)}{M(\cdot)}. \qquad (54)$$

Now, M'/M is positive since $\gamma^*>0$.[7] Therefore,

$$\frac{\gamma^*}{\bar{S}}+\frac{r-r^*}{(r-r^*)\bar{S}-\phi X(t)}>0, \qquad (55)$$

that is[8]

---

[7] Using the property $M'(a,b,x)=\dfrac{a}{b}M(a+1,b+1,x)$ of the confluent hypergeometric function and the fact that $M(a,b,x)=e^x M(b-a,b,-x)$ yields

$$\frac{M'\left(\gamma^*,2+\gamma^*,-\dfrac{2\phi X(t)}{\sigma^2\bar{S}}\right)}{M\left(\gamma^*,2+\gamma^*,-\dfrac{2\phi X(t)}{\sigma^2\bar{S}}\right)}=\frac{\gamma^*}{2+\gamma^*}\frac{M\left(1+\gamma^*,3+\gamma^*,-\dfrac{2\phi X(t)}{\sigma^2\bar{S}}\right)}{M\left(\gamma^*,2+\gamma^*,-\dfrac{2\phi X(t)}{\sigma^2\bar{S}}\right)}$$

$$=\frac{\gamma^*}{2+\gamma^*}\frac{M\left(2,3+\gamma^*,\dfrac{2\phi X(t)}{\sigma^2\bar{S}}\right)}{M\left(2,2+\gamma^*,\dfrac{2\phi X(t)}{\sigma^2\bar{S}}\right)}>0.$$

$$\bar{S} < \frac{\phi X(t)}{r - r^*} \frac{\gamma^*}{1 + \gamma^*}, \tag{56}$$

which means that, when premium payments can be made from existing assets, the optimal bankruptcy trigger is lower than in the case where the premium is paid by the shareholders.[9] One might suspect that the equity holders' optimal bankruptcy strategy is to wait until the asset value reaches zero before defaulting on insurance premium payments. The reason is that when the insurance premium is paid from bank assets, paying it costs nothing directly to equity holders. Therefore, the payment has no commitment value and the incentive scheme breaks down. In fact, one can show that the first-order condition (54) will be satisfied asymptotically as $\bar{S} \downarrow 0$.[10] Figure 6.7 illustrates this result graphically.

---

[8] One can check by contradiction that $(r - r^*)\bar{S} - \phi X(t)$ must be negative in the optimum: suppose that this were not the case. Then, we would have

$$\bar{S} > \frac{\phi X(t)}{r - r^*} \frac{\gamma^*}{1 + \gamma^*}.$$

Together, these conditions would imply

$$\left(\frac{\gamma^*}{\bar{S}} + \frac{r - r^*}{(r - r^*)\bar{S} - \phi X(t)} \right) \frac{\sigma^2 \bar{S}^2}{2\phi X(t)} > \frac{\gamma^*}{\bar{S}} \frac{\sigma^2 \bar{S}^2}{2\phi X(t)} = \gamma^* \frac{\sigma^2 \bar{S}}{2\phi X(t)}$$

$$> \gamma^* \frac{\sigma^2}{2\phi X(t)} \frac{\phi X(t)}{r - r^*} \frac{\gamma^*}{1 + \gamma^*} = \frac{\gamma^*}{1 + \gamma^*} > \frac{\gamma^*}{2 + \gamma^*}$$

$$> \frac{\gamma^*}{2 + \gamma^*} \frac{M\left(2,3 + \gamma^*, \frac{2\phi X(t)}{\sigma^2 \bar{S}}\right)}{M\left(2,2 + \gamma^*, \frac{2\phi X(t)}{\sigma^2 \bar{S}}\right)} = \frac{M'\left(\gamma^*, 2 + \gamma^*, -\frac{2\phi X(t)}{\sigma^2 \bar{S}}\right)}{M\left(\gamma^*, 2 + \gamma^*, -\frac{2\phi X(t)}{\sigma^2 \bar{S}}\right)},$$

thus violating the optimality condition (54).

[9] Remember that, when the premium is paid by the shareholders, the analysis of Chapter 3 applies and the bankruptcy-triggering asset value is given by

$$\bar{S} = \frac{\phi X(t)}{r - r^*} \frac{\gamma^*}{1 + \gamma^*}.$$

[10] From Slater (1968),

$$\lim_{x \uparrow \infty} \frac{M(a,b,x)}{e^x x^{a-b}} = \frac{\Gamma(b)}{\Gamma(a)},$$

where $\Gamma(a)$ is the gamma function, $\Gamma(a) = \int_0^\infty s^{a-1} e^{-s} ds$. Applying this property to the right hand side of (54) using the relationship

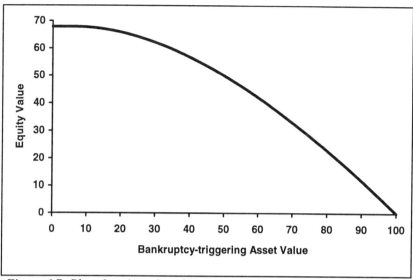

Figure 6.7: *Plot of equity value (51) against the bankruptcy-triggering asset value $\overline{S}$ for the following parameter values: $\phi = 0.01$, $S = 100$, $X = 50$, $r - r^* = 0.01$ and $\sigma = 0.2$. When the insurance premium can be paid from bank assets, it has no commitment value and the equity holders' optimal bankruptcy strategy is $\overline{S} = 0$.*

$$\frac{M'\left(\gamma^*, 2+\gamma^*, -\dfrac{2\phi X(t)}{\sigma^2 \overline{S}}\right)}{M\left(\gamma^*, 2+\gamma^*, -\dfrac{2\phi X(t)}{\sigma^2 \overline{S}}\right)} = \frac{\gamma^*}{2+\gamma^*} \frac{M\left(2,3+\gamma^*, \dfrac{2\phi X(t)}{\sigma^2 \overline{S}}\right)}{M\left(2,2+\gamma^*, \dfrac{2\phi X(t)}{\sigma^2 \overline{S}}\right)}$$

demonstrated in Footnote 7 yields

$$\lim_{\overline{S}\downarrow 0} \frac{M'\left(\gamma^*, 2+\gamma^*, -\dfrac{2\phi X(t)}{\sigma^2 \overline{S}}\right)}{M\left(\gamma^*, 2+\gamma^*, -\dfrac{2\phi X(t)}{\sigma^2 \overline{S}}\right)} = \frac{\gamma^*}{2+\gamma^*} \lim_{\overline{S}\downarrow 0} \frac{M\left(2,3+\gamma^*, \dfrac{2\phi X(t)}{\sigma^2 \overline{S}}\right)}{M\left(2,2+\gamma^*, \dfrac{2\phi X(t)}{\sigma^2 \overline{S}}\right)}$$

$$= \frac{\gamma^*}{2+\gamma^*} \lim_{x\uparrow\infty} \frac{M(2,3+\gamma^*,x)}{M(2,2+\gamma^*,x)} = \frac{\gamma^*}{2+\gamma^*} \lim_{x\uparrow\infty} \frac{\Gamma(3+\gamma^*)e^x x^{-1-\gamma^*}/\Gamma(2)}{\Gamma(2+\gamma^*)e^x x^{-\gamma^*}/\Gamma(2)}$$

$$= \frac{\gamma^*}{2+\gamma^*} \frac{\Gamma(3+\gamma^*)}{\Gamma(2+\gamma^*)} \lim_{x\uparrow\infty}\frac{1}{x} = \gamma^* \lim_{x\uparrow\infty}\frac{1}{x} = 0,$$

so the right hand side of (54) is zero. Straightforward calculations then show that the left hand side of (54) is also zero, so that (54) is asymptotically satisfied as $\overline{S}\downarrow 0$.

6.7.2 Merton's Solution

In a model of surplus insurance, Merton (1997) constructs an incentive contract to lead the insiders to reveal the true current asset value and declare bankruptcy as soon as the value of the surplus falls to zero. In this section, a similar analysis is conducted for the case of deposit insurance. The basic intuition is that the guarantor can lead the bank to declare bankruptcy early by promising to pay a certain amount to equity holders if they choose to liquidate the bank. This promise can be interpreted as an incentive contract.

For simplicity, suppose that no insurance premium is paid and that the value of the bank's assets follows the usual geometric Brownian motion

$$dS = \mu S dt + \sigma S d\tilde{z} . \tag{57}$$

As before, the deposits are assumed to follow

$$X(t) = X_0 \cdot e^{r^*t} . \tag{58}$$

Let $E(S)$ denote the value of bank equity, and ignore bank capital and liquidation costs, which would only complicate the analysis without changing the main insight. Making the change in variables

$$V = \frac{S_t}{X(t)} \tag{59}$$

and defining

$$F(V) = \frac{E(S)}{X(t)} , \tag{60}$$

F satisfies the following ordinary differential equation:

$$\tfrac{1}{2}\sigma^2 V^2 F'' + (r - r^*)VF' - (r - r^*)F = 0 . \tag{61}$$

The general solution is

$$F(V) = \alpha_1 V + \alpha_2 V^{-\gamma^*}, \quad \gamma^* \equiv 2\frac{r - r^*}{\sigma^2} . \tag{62}$$

Now, suppose the guarantor wants the assets to be turned in as soon as $V = c$, that is, wants to render

$$\overline{V} = c \tag{63}$$

self-enforcing, where $\overline{V}$ denotes the assets-to-deposits ratio at which the equity holders choose to turn the bank's assets to the guarantor for liquidation. To achieve this goal, the guarantor can enter the following contract with the bank: if the bank turns the assets to the guarantor, it will receive

$$Max[0; a + b\overline{V}] . \tag{64}$$

Applying this boundary condition to equation (62) with $\alpha_1 = 1$ to avoid arbitrage yields

$$F(\overline{V}) = \overline{V} + \alpha_2 \overline{V}^{-\gamma^*} = a + b\overline{V} \quad \Leftrightarrow \quad \alpha_2 = \left(a + (b-1)\overline{V}\right)\overline{V}^{\gamma^*}. \quad (65)$$

F therefore becomes

$$F(V) = V + \left(a + (b-1)\overline{V}\right)\left(\frac{V}{\overline{V}}\right)^{-\gamma^*}, \quad (66)$$

and the value of equity equals[11]

$$E(S) = X\left(V + \left(a + (b-1)\overline{V}\right)\left(\frac{V}{\overline{V}}\right)^{-\gamma^*}\right). \quad (67)$$

Bank equity holders choose that value of $\overline{V}$ which maximizes the value of equity, that is, set

$$\frac{\partial E(S)}{\partial \overline{V}} = X\left((b-1)\left(\frac{V}{\overline{V}}\right)^{-\gamma^*} + \frac{\gamma^*}{\overline{V}}\left(a + (b-1)\overline{V}\right)\left(\frac{V}{\overline{V}}\right)^{-\gamma^*}\right)$$

$$= XV^{-\gamma^*}\left((1+\gamma^*)(b-1)\overline{V}^{\gamma^*} + \gamma^* a\overline{V}^{\gamma^*-1}\right) = 0, \quad (68)$$

thus yielding[12]

$$\overline{V} = \frac{a}{1-b}\frac{\gamma^*}{1+\gamma^*}. \quad (69)$$

The guarantor, knowing that the bank will choose (69), should therefore set a and b such that

$$\overline{V} = \frac{a}{1-b}\frac{\gamma^*}{1+\gamma^*} = c, \quad (70)$$

which will be self-enforcing if

---

[11] For convenience, equity value is written as a function of $V = S_t / X(t)$.

[12] Since

$$\left.\frac{\partial^2 E(S)}{\partial \overline{V}^2}\right|_{\overline{V} = \frac{a}{1-b}\frac{\gamma^*}{1+\gamma^*}} = XV^{-\gamma^*}\left((1+\gamma^*)(b-1)\overline{V}^{\gamma^*-1}\gamma^* + \gamma^*(\gamma^*-1)a\overline{V}^{\gamma^*-2}\right)$$

$$= XV^{-\gamma^*}\overline{V}^{\gamma^*-2}\gamma^*\left((1+\gamma^*)(b-1)\overline{V} + (\gamma^*-1)a\right)$$

$$= XV^{-\gamma^*}\overline{V}^{\gamma^*-2}\gamma^*\left(-a\gamma^* + (\gamma^*-1)a\right)$$

$$= -a\gamma^* XV^{-\gamma^*}\overline{V}^{\gamma^*-2} < 0,$$

it is a maximum.

$$a + b\overline{V} = a + b\frac{a}{1-b}\frac{\gamma *}{1+\gamma *}$$

$$= \frac{a\big((1-b)(1+\gamma *) + b\gamma *\big)}{(1-b)(1+\gamma *)} = \frac{a\big(1+\gamma * -b\big)}{(1-b)(1+\gamma *)} > 0. \tag{71}$$

This means that, in order to induce the bank to tender assets as the boundary (70) is reached, the guarantor has to *pay* a positive amount to equity holders when they turn in their assets for liquidation. This is exactly the opposite of what we tried to do in the previous section, where the bank's no more paying the insurance premium was to trigger liquidation.

A natural question that arises is that of knowing how contract (70) changes the *risk-taking* incentives of the bank. Substituting the equity holders' optimal bankruptcy decision (70) into the equity value (67) yields

$$E(S) = X\left[V + \left(a + (b-1)\frac{a}{1-b}\frac{\gamma *}{1+\gamma *} \right)\left(\frac{V}{\dfrac{a}{1-b}\dfrac{\gamma *}{1+\gamma *}} \right)^{-\gamma *} \right] \tag{72}$$

$$= X\left[V + \frac{a}{1+\gamma *}\left(\frac{(1-b)(1+\gamma *)V}{a\gamma *} \right)^{-\gamma *} \right].$$

Removing constant terms and taking logs yields

$$\Phi = \ln a - \ln(1+\gamma *) - \gamma * \ln\left(\frac{(1-b)(1+\gamma *)V}{a\gamma *} \right), \tag{73}$$

which is to be maximized. Now,

$$\frac{\partial \Phi}{\partial \gamma *} = -\frac{1}{1+\gamma *} - \ln\left(\frac{(1-b)(1+\gamma *)V}{a\gamma *} \right) - \gamma *\left(\frac{1}{1+\gamma *} - \frac{1}{\gamma *} \right)$$

$$= -\left(\frac{1}{1+\gamma *} + \ln\left(\frac{(1-b)(1+\gamma *)V}{a\gamma *} \right) + \frac{\gamma *}{1+\gamma *} - 1 \right) \tag{74}$$

$$= -\ln\left(\frac{(1-b)(1+\gamma *)V}{a\gamma *} \right) < 0,$$

since $\ln\big((1-b)(1+\gamma *)V / (a\gamma *)\big)$ is positive by assumption. Hence, using

$$\frac{d\gamma *}{d\sigma^2} = -\frac{\gamma *}{\sigma^2} < 0, \tag{75}$$

we get

$$\frac{\partial \Phi}{\partial \sigma^2} = \frac{\partial \Phi}{\partial \gamma *}\frac{d\gamma *}{d\sigma^2} > 0, \tag{76}$$

which implies that under contract (70), the bank has an incentive to increase asset risk in order to raise equity value. This clearly restricts the usefulness of contract (70) for the practice of deposit insurance. At this point, it is interesting to note that we are facing a situation that is very similar to that encountered in Chapter 3. When asset value is observable, the threat of liquidation by the guarantor leads the bank to reduce its risk as much as possible. However, when asset value is unobservable, the guarantor's attempt to construct an incentive contract triggering early liquidation fails because it gives the bank an incentive to increase its asset risk.

6.8 Conclusion

Deposit insurance prevents bank runs by ensuring depositors that they will not suffer if others choose to withdraw their money and they do not. Building on the results of Chapter 5, and assuming that the bank's asset value is perfectly observable and that the guarantor can seize the bank's assets and liquidate them immediately if he wishes, this chapter began by valuing deposit insurance as a perpetual put option and bank equity as a knock-out perpetual call option, conditional on the guarantor's liquidation strategy.

In order to analyze the costs and benefits of deposit insurance, a measure of social welfare, the difference between the bank's equity value and the value of the insurance guarantee, was introduced. It was shown that deposit insurance will be socially beneficial if it can avoid costly fire sales or asset liquidation altogether.

In analyzing the *incentive* effects of deposit insurance, it was demonstrated that risk-reduction incentives similar to those implied by the possibility of bank runs exist if the insurer can monitor asset value perfectly and liquidate assets immediately. Moreover, the guarantor's forbearance was shown to lower the bank's optimal capital share. Deposit insurance can therefore be considered as a substitute for bank capital.

However, the bank's incentive to reduce asset risk is not robust. When the assumptions of immediate liquidation and perfect observability of current asset value by the guarantor are relaxed, bank behavior changes. More specifically, if there are *liquidation delays*, risk-shifting behavior by banks may arise once liquidation has been announced by the guarantor.

If the guarantor is unable to *observe* asset value altogether, then an incentive contract such that the bank tenders its assets as soon as they reach some specific value can be constructed. This contract, however, gives rise to a risk-shifting problem. This limits its usefulness for the practice of deposit insurance. It also stresses the importance of a close supervision of insured depository institutions in order to avoid the incentive problems resulting from unobservable asset values.

In this setting, we are thus facing a situation that is very similar to that encountered in Chapter 3. When asset value is observable, the threat of liquidation by the guarantor leads the bank to reduce its risk as much as possible. However, when asset value is unobservable, the guarantor's attempt to construct an incentive contract triggering early liquidation might fail because it gives the bank an incentive to increase its asset risk. Again, monitoring asset value and monitoring asset risk can be considered as substitutes.

7. Summary and Conclusions

In recent years, game theory has faced methodological problems in handling uncertainty and timing decisions in dynamic models. This book presents a method to analyze these kind of situations, the *game theory analysis of options,* which can be understood as an attempt to integrate game theory and option pricing.

As it is presented in Chapter 1, the game theory analysis of options in effect replaces the maximization of *expected utility* encountered in classical game theory models with the maximization of the value of an *option*, which gives the arbitrage-free value of the payoffs to the players and can therefore be considered as a proxy for expected utility. Over the expected-utility approach, the option-pricing approach has the advantage that it automatically takes the time value of money and the price of *risk* into account. The main advantage of the method, however, lies in its ability to *separate* two issues in economic model building, namely, that of the valuation of uncertain future payoffs and that of strategic interactions. Using option pricing, arbitrage-free values for the payoffs to the economic agents can be obtained. These values are then inserted into the strategic games between the agents, which can thus be analyzed more realistically. By integrating game theory and option pricing, the game theory analysis of options actually provides the link between *markets* and *organizations:* while the use of option pricing enables the valuation of the players' payoffs using market criteria, game theory modeling takes the institutional structure of organizations into account.

In the subsequent chapters, a number of examples from the theory of corporate finance and financial intermediation in continuous time are presented.

Chapter 2 demonstrates that the two classical problems in financial contracting, namely, the risk-shifting problem and the observability problem, are very closely related. Whereas solving the risk-shifting problem calls for a *linear* profit-sharing rule between the lender and the borrower, solving the observability problem results in a concave payoff to the lender and a *convex* payoff to the borrower. This convexity, however, gives the borrower an incentive to increase project risk. There is, therefore, a tradeoff between both problems. They can only be solved simultaneously if the lender's claim becomes risk-free, which can be achieved through full collateralization of the loan.

The analysis of endogenous bankruptcy in Chapter 3 demonstrates that the amount of debt and the interest rates on this debt influence the equity holders' bankruptcy decision. This, in turn, has an effect on the structure of debt contracts and the capital structure choice at initial time. In particular, the equity holders choose a capital structure so as to hold the expected life of companies constant. Finally, Chapter 3 shows how debt holders can use interest payments to lead the equity holders to declare bankruptcy as soon as the asset value reaches a pre-specified level.

The analysis of junior debt in Chapter 4 demonstrates that a junior debt issue reduces the value of senior debt through its influence on the equity holders' bankruptcy decision. This wealth transfer between different classes of security holders distorts the equity holders' capital structure choice and may give rise to socially suboptimal capital structures. Moreover, this result invalidates the conventional wisdom that seniority fully protects debt holders against adverse wealth effects resulting from subsequent debt issues.

The analysis of bank runs in Chapter 5 shows that bank runs may occur as soon as the value of the bank's assets (net of liquidation costs) falls below the face value of deposits. Modeling the value of bank equity as a knock-out call option, it is demonstrated that the possibility of a bank run leads the bank to reduce its asset risk, and eventually to invest everything in the risk-free asset. Demandable debt can thus be understood as an optimal contractual arrangement to preclude banks from engaging in risk-shifting activities. It results in a *separation* of the returns for the time value of money and for the riskiness of the underlying venture.

Using the results of Chapter 5, Chapter 6 first analyzes the costs and benefits of deposit insurance. In particular, it shows that deposit insurance is socially beneficial to the extent that it lowers liquidation costs. This occurs through the avoidance of costly fire-sales that would be triggered in the event of a bank run. Liquidation will occur less frequently if a fall in the asset value below the face value of deposits is tolerated by the guarantor. Consequently, some forbearance on the part of the guarantor typically is socially beneficial. In analyzing the incentive effects of deposit insurance, it is shown that its existence has no influence on the bank's investment incentives as long as the guarantor can monitor the asset value perfectly and is able to seize or liquidate bank assets immediately. If this is not the case, however, some interesting incentive problems may arise. More precisely, if the guarantor has to wait before he can liquidate the bank, his announcement to liquidate will induce the bank to increase its risk. If the guarantor cannot observe asset value at all, then he has to construct an incentive contract leading the bank to liquidate on its own in due time. The analysis demonstrates that the incentive scheme developed in Chapter 3 might break down. Instead, the guarantor has to enter a contract with the bank promising to pay a certain positive amount to the shareholders if they declare bankruptcy. This contract, however, gives rise to a risk-shifting problem. Hence, monitoring asset value and monitoring asset risk can be considered as substitutes.

This short overview of the main results of this text illustrate how powerful an instrument the game theory analysis of options is. The methodology presented in the preceding chapters can, no doubt, be applied to other problems of dynamic strategic interactions under uncertainty, such as real investment.

The method has, however, some important limitations. First, although separation of valuation and strategic issues is possible, the mathematical expressions obtained in the models of the preceding chapters were, in general, quite complex. This mathematical complexity, which is inherent to option pricing, means that simple, closed-form solutions might not always be obtainable.

Second, the method only works easily if the players' optimal strategies are non-stochastic, i.e. do not depend on the value taken by the state variable. The reason is that if the optimal strategy is state- and path-dependent, valuation of the players' payoffs using traditional option · pricing methodologies becomes tedious, if not impossible. As a result, the players' optimal strategies in the preceding stages cannot be computed. For instance, in the analysis of junior debt, it is not possible to value senior debt *before* junior debt is issued since the equity holders' decision to issue junior debt depends on current project value, which is stochastic. These restrictions should be kept in mind when applying this method. They also provide directions for further research.

Finally, it should be remembered that continuous-time modeling is an abstraction of reality. Thus, caution is required when interpreting the results obtained with the method. The analysis of any model mostly requires restrictive assumptions. Nevertheless, continuous-time analysis can provide a good approximation.

Table of Symbols

Chapter 1

a	Payout to the holders of the underlying asset per unit time
A	Strategy of player I
$\overline{A}(S)$	Optimal strategy of player I
b	Payout to the holders of the contingent claim per unit time
$\overline{B}(A,S)$	Optimal strategy of player II
B	Strategy of player II
$d\tilde{z}$	Increment of a standard Wiener process
$f(x,\overline{x})$	Function
$F(S,t)$	Contingent claim value
$G(A,B,S)$	Payoff to player I as given by option pricing
$H(A,B,S)$	Payoff to player II as given by option pricing
$P_\infty(S)$	Perpetual put option value
r	Risk-free interest rate
S	Underlying asset value
$\overline{S}$	Perpetual put option holder's exercise strategy
t	Time
X	Exercise price of the perpetual put option
γ	$\equiv 2r/\sigma^2$
μ	Drift
σ	Instantaneous standard deviation

Chapter 2

c	Verification cost
$C(X_2)$	Value of a call option with exercise price X_2
D	Lump sum
D_0	Initial outside financing provided by the lender
$d\tilde{z}$	Increment of a standard Wiener process
E_0	Initial equity financing provided by the borrower
$f(\bar{S})$	Promised payment to the lender at maturity
$g(S)$	Distribution of gross project return
N	Cumulative standard normal distribution function
$P(X_1)$	Value of a put option with exercise price X_1
r	Risk-free interest rate
S	Project value
S_0	Initial project value
S_{ex}	Current value of a claim on S without the right to dividends
$\bar{S}$	Terminal project value
T	Total project life
X	Collateral amount
α	Number of put options
$[\alpha,\beta]$	Support of the distribution of gross project return $g(S)$
$[\alpha,\gamma)$	Verification region
β	Number of call options
β,β'	Percentage of financing received from the lender
δ	Payout rate
μ	Drift
Π	Value of the payment to the lender as given by option pricing
σ	Instantaneous standard deviation
τ	Remaining project life
ξ	Probability that verification occurs if the good state is announced
Ψ	Must be constant for a contract to be dynamically stable

Chapter 3

C	Agency cost of debt
$D(t)$	Face value of debt at time t
$\overline{D}(t)$	Optimal face value of debt at time t
$\overline{D}_0$	Optimal face value of debt at initial time
$d\tilde{z}$	Increment of a standard Wiener process
$E(S)$	Equity value
$F(S)$	Market value of debt
$G(V)$	Normalized market value of debt $G(V) = F(S)/D(t)$
I	Initial investment by the shareholders
$K(S)$	Value of bankruptcy costs
r	Risk-free interest rate
r^*	Growth rate of the face value of debt $D(t)$
S	Firm asset value
S_0	Initial asset value
S_B	Bankruptcy-triggering asset value
$\overline{S}_B$	Bankruptcy-triggering asset value as specified by a covenant
V	Normalized asset value $V = S/D(t)$
$TB(S)$	Value of the tax benefits
w	Scale-up factor
$W(S)$	Firm value
x	Variable following an arithmetic Brownian motion
x_0	Initial value for x
α	Proportional bankruptcy cost
β	Payout rate
γ^*	$\equiv 2(r-r^*)/\sigma^2$
θ	Tax rate
ϑ	$\equiv (1-\theta)/\theta$
$\lambda_{1,2}$	Roots of the characteristic equation
μ	Drift
σ	Instantaneous standard deviation
$\overline{\tau}$	Mean time to bankruptcy
ϕ	Interest rate to be effectively paid on debt
$\overline{\phi}$	Interest rate to be paid on debt in the incentive contract

Chapter 4

D_1, D_2	Face value of senior and junior debt, respectively
$d\tilde{z}$	Increment of a standard Wiener process
$E(S)$	Equity value
F_1, F_2	Market value of senior and junior debt, respectively
$K(S)$	Value of bankruptcy costs
r	Risk-free interest rate
S	Firm asset value
S_B	Bankruptcy-triggering asset value
$TB(S)$	Value of the tax benefits
$W(S)$	Firm value
α	Proportional bankruptcy cost
γ	$\equiv 2r/\sigma^2$
θ	Tax rate
μ	Drift
σ	Instantaneous standard deviation
ϕ_1, ϕ_2	Interest rate on senior and junior debt, respectively

Chapter 5

$B(t)$	Value of the amount invested in the risk-free asset at time t
B_0	Amount initially invested in the risk-free asset
C_∞	Value of bank equity, modeled as a perpetual knock-out call option
$d\tilde{z}$	Increment of a standard Wiener process
$F(V)$	Normalized equity value $F(V) = C_\infty / X(t)$
G	Expected profit from funding a bank
$K(t)$	Knock-out asset value at time t $K(t) = (1-\beta)X(t)/(1-\alpha)$
L	Present value of the payoff to the equity holders if they never liquidate
$L(t)$	Payoff to the equity holders if they liquidate at time t
$L_0(t)$	Present value of the payoff to the equity holders if they liquidate at time t
r	Risk-free interest rate
$r*$	Interest rate paid on deposits
S	Bank asset value
S_0	Initial asset value
x	Capital provided by equity holders at initial time
$\bar{x}$	Optimal capital to be provided by equity holders at initial time
$X(t)$	Face value of deposits at time t
X_0	Amount deposited initially
V	Normalized asset value $V = (1-\beta)(1+x)S/X(t)$
w	Scale-up factor
α	Liquidation cost in the event of a run
β	Liquidation cost if no run occurs
$\gamma *$	$\equiv 2(r-r*)/\sigma^2$
μ	Drift
σ	Instantaneous standard deviation

Chapter 6

a, b	Incentive contract parameters
c	Target liquidation-triggering assets-to-deposits ratio $\overline{V}$
C_∞, C_∞^-	Value of bank equity, modeled as a perpetual knock-out call option, with and without deposit insurance, respectively
$d\tilde{z}$	Increment of a standard Wiener process
$E(S)$	Bank equity value when asset value is unobservable
$F(V)$	Normalized derivative security value $F(V) = P_\infty / X(t)$, $F(V) = C_\infty / X(t)$ or $F(V) = E(S)/X(t)$
G	Expected profit from funding a bank
$h(Z)$	$\equiv F(V)Z^{-\gamma^*}e^Z$
M	Confluent hypergeometric function
P_∞	Value of the deposit insurance guarantee, modeled as a put option
r	Risk-free interest rate
r^*	Interest rate paid on deposits
S	Bank asset value
S_0	Initial asset value
$\overline{S}$	Liquidation-triggering asset value
x	Capital provided by equity holders at initial time
$X(t)$	Face value of deposits at time t
X_0	Amount deposited initially
V	Assets-to-deposits ratio $V = (1-\beta)(1+x)S/X(t)$ or $V = S/X(t)$
$\overline{V}$	Liquidation-triggering assets-to-deposits ratio
Z	$\equiv 2\phi/(\sigma^2 V)$
α	Liquidation cost in the event of a run
β	Liquidation cost if the guarantor closes the bank
γ^*	$\equiv 2(r-r^*)/\sigma^2$
Γ	Gamma function
μ	Drift
Π	Value of the social surplus
σ	Instantaneous standard deviation
τ	Liquidation delay
ϕ	Instantaneous insurance premium

References

Baer, H., Brewer, E. (1986): Uninsured Deposits as a Source of Market Discipline. Federal Reserve Bank of Chicago *Economic Perspectives* (November/December), 25-31

Black, F., Cox, J. C. (1976): Valuing Corporate Securities: Some Effects of Bond Indenture Provisions. *Journal of Finance* 31, 351-367

Black, F., Scholes, M. S. (1973): The Pricing of Options and Corporate Liabilities. *Journal of Political Economy* 81, 637-659

Calomiris, C. W. (1997): Banks and Banking: Function, Structure, and Development, *unpublished*

Calomiris, C. W., Kahn, C. M. (1991): The Role of Demandable Debt in Structuring Optimal Banking Arrangements. *American Economic Review* 81, 497-513

Chesney, M., Gibson, R. (1994): The Investment Policy and the Pricing of Equity in a Levered Firm: A Reexamination of the Contingent Claims' Valuation Approach. Working Paper 9403, Institute of Banking and Financial Management, University of Lausanne

Diamond, D. W., Dybvig, P. H. (1983): Bank Runs, Deposit Insurance, and Liquidity. *Journal of Political Economy* 91, 401-419

Fan, H., Sundaresan, S. (1997): Debt Valuation, Strategic Debt Service and Optimal Dividend Policy. Working Paper, Columbia University

Fudenberg, D., Tirole, J. (1991): *Game Theory.* The MIT Press, Cambridge MA

Gale, D., Hellwig, M. (1985): Incentive-Compatible Debt Contracts: The One-Period Problem. *Review of Economic Studies* 52, 647-663

Gibbons, R. (1992): *A Primer in Game Theory.* Harvester-Wheatsheaf, Hempstead, UK.

Hart, O., Moore, J. (1995): Debt and Seniority: An Analysis of the Role of Hard Claims in Constraining Management. *American Economic Review* 85, 567-585

Huang, C. (1985): Information Structure and Equilibrium Asset Prices. *Journal of Economic Theory* 34, 33-71

Huang, C. (1987): An Intertemporal General Equilibrium Asset Pricing Model: The Case of Diffusion Information. *Econometrica* 55, 117-142

Hull, J. C. (1993): *Options, Futures and Other Derivative Securities,* 2nd Edition. Prentice-Hall, Englewood Cliffs NJ

Ingersoll, J. E. (1987): *Theory of Financial Decision Making.* Rowman & Littlefield, Savage MD

Jensen, M. C. (1986): Agency Costs of Free Cash Flow, Corporate Finance, and Takeovers. *American Economic Review* 76, 323-329

Jensen, M. C., Meckling, W. H. (1976): Theory of the Firm: Managerial Behavior, Agency Costs and Ownership Structure. *Journal of Financial Economics* 3, 305-360

John, K., John, T. A., Senbet, L. W. (1991): Risk-Shifting Incentives of Depository Institutions: A new Perspective on Federal Deposit Insurance Reform. *Journal of Banking and Finance* 15, 895-915

Kane, E. J. (1995): Three Paradigms for the Role of Capitalization Requirements in Insured Financial Institutions. *Journal of Banking and Finance* 19, 431-459

Kaufman, G. G. (1988): Bank Runs: Causes, Benefits and Costs. *Cato Journal* 7, 559-587

Leland, H. E. (1994): Corporate Debt Value, Bond Covenants, and Optimal Capital Structure. *Journal of Finance* 49, 1213-1252

Longstaff, F. A. (1990): Pricing Options with Extendible Maturities: Analysis and Applications. *Journal of Finance* 45, 935-957

Mello, A. S., Parsons, J. E. (1992): Measuring the Agency Cost of Debt. *Journal of Finance* 47, 1887-1904

Merton, R. C. (1973): Theory of Rational Option Pricing. *Bell Journal of Economics and Management Science* 4, 141-183. Reproduced as Chapter 8 of Merton (1990)

Merton, R. C. (1977): On the Pricing of Contingent Claims and the Modigliani-Miller Theorem. *Journal of Financial Economics* 5, 241-249. Reproduced as Chapter 12 of Merton (1990)

Merton, R. C. (1989): On the Application of the Continuous-Time Theory of Finance to Financial Intermediation and Insurance. *Geneva Papers on Risk and Insurance* 14, 225-261. Reproduced as Chapter 14 of Merton (1990)

Merton, R. C. (1990): *Continuous Time Finance*, 2nd Edition. Blackwell, Cambridge MA

Merton, R. C. (1997): A Model of Contract Guarantees for Credit-Sensitive, Opaque Financial Intermediaries. Working Paper 97-091, Harvard Business School

Müller, H. M. (1997): *The Theory of Moral Hazard*. Ph.D. Dissertation, University of St. Gallen

Myers, S. C. (1977): Determinants of Corporate Borrowing. *Journal of Financial Economics* 5, 147-175

Perotti, E. C., Spier, K. E. (1993): Capital Structure as a Bargaining Tool: The Role of Leverage in Contract Renegociation. *American Economic Review* 83, 1131-1141

Postlewaite, A. and Vives, X. (1987): Bank Runs as an Equilibrium Phenomenon. *Journal of Political Economy* 95, 485-491

Simon, C. P. and Blume, L. (1994): *Mathematics for Economists*. Norton, New York

Seward, J. K. (1990): Corporate Financial Policy and the Theory of Financial Intermediation. *Journal of Finance* 45, 351-377

Slater, L. J. (1968): *Confluent Hypergeometric Functions*. Chapter 13 in: Abramowitz, M., Stegun, I. A. (Eds.): *Handbook of Mathematical Functions*. Applied Mathematics Series 55, U.S. National Bureau of Standards, Washington D.C.

Stiglitz, J. E. and Weiss, A. (1981): Credit Rationing in Markets with Imperfect information. *American Economic Review* 71, 393-410

Townsend, R. M. (1979): Optimal Contracts and Competitive Markets with Costly State Verification. *Journal of Economic Theory* 21, 265-293

von Neumann, J. (1928): Zur Theorie der Gesellschaftsspiele. *Mathematische Annalen* 100, 295-320

Printing: Druckhaus Beltz, Hemsbach
Binding: Buchbinderei Schäffer, Grünstadt

cture Notes in Economics
nd Mathematical Systems

For information about Vols. 1–284
please contact your bookseller or Springer-Verlag

Vol. 322: T. Kollintzas (Ed.), The Rational Expectations Equilibrium Inventory Model. XI, 269 pages. 1989.

Vol. 323: M.B.M. de Koster, Capacity Oriented Analysis and Design of Production Systems. XII, 245 pages. 1989.

Vol. 324: I.M. Bomze, B.M. Pötscher, Game Theoretical Foundations of Evolutionary Stability. VI, 145 pages. 1989.

Vol. 325: P. Ferri, E. Greenberg, The Labor Market and Business Cycle Theories. X, 183 pages. 1989.

Vol. 326: Ch. Sauer, Alternative Theories of Output, Unemployment, and Inflation in Germany: 1960–1985. XIII, 206 pages. 1989.

Vol. 327: M. Tawada, Production Structure and International Trade. V, 132 pages. 1989.

Vol. 328: W. Güth, B. Kalkofen, Unique Solutions for Strategic Games. VII, 200 pages. 1989.

Vol. 329: G. Tillmann, Equity, Incentives, and Taxation. VI, 132 pages. 1989.

Vol. 330: P.M. Kort, Optimal Dynamic Investment Policies of a Value Maximizing Firm. VII, 185 pages. 1989.

Vol. 331: A. Lewandowski, A.P. Wierzbicki (Eds.), Aspiration Based Decision Support Systems. X, 400 pages. 1989.

Vol. 332: T.R. Gulledge, Jr., L.A. Litteral (Eds.), Cost Analysis Applications of Economics and Operations Research. Proceedings. VII, 422 pages. 1989.

Vol. 333: N. Dellaert, Production to Order. VII, 158 pages. 1989.

Vol. 334: H.-W. Lorenz, Nonlinear Dynamical Economics and Chaotic Motion. XI, 248 pages. 1989.

Vol. 335: A.G. Lockett, G. Islei (Eds.), Improving Decision Making in Organisations. Proceedings. IX, 606 pages. 1989.

Vol. 336: T. Puu, Nonlinear Economic Dynamics. VII, 119 pages. 1989.

Vol. 337: A. Lewandowski, I. Stanchev (Eds.), Methodology and Software for Interactive Decision Support. VIII, 309 pages. 1989.

Vol. 338: J.K. Ho, R.P. Sundarraj, DECOMP: An Implementation of Dantzig-Wolfe Decomposition for Linear Programming. VI, 206 pages.

Vol. 339: J. Terceiro Lomba, Estimation of Dynamic Econometric Models with Errors in Variables. VIII, 116 pages. 1990.

Vol. 340: T. Vasko, R. Ayres, L. Fontvieille (Eds.), Life Cycles and Long Waves. XIV, 293 pages. 1990.

Vol. 341: G.R. Uhlich, Descriptive Theories of Bargaining. IX, 165 pages. 1990.

Vol. 342: K. Okuguchi, F. Szidarovszky, The Theory of Oligopoly with Multi-Product Firms. V, 167 pages. 1990.

Vol. 343: C. Chiarella, The Elements of a Nonlinear Theory of Economic Dynamics. IX, 149 pages. 1990.

Vol. 344: K. Neumann, Stochastic Project Networks. XI, 237 pages. 1990.

Vol. 345: A. Cambini, E. Castagnoli, L. Martein, P Mazzoleni, S. Schaible (Eds.), Generalized Convexity and Fractional Programming with Economic Applications. Proceedings, 1988. VII, 361 pages. 1990.

Vol. 346: R. von Randow (Ed.), Integer Programmin Related Areas. A Classified Bibliography 1984–1987. 514 pages. 1990.

Vol. 347: D. Ríos Insua, Sensitivity Analysis in Mui objective Decision Making. XI, 193 pages. 1990.

Vol. 348: H. Störmer, Binary Functions and their Applications. VIII, 151 pages. 1990.

Vol. 349: G.A. Pfann, Dynamic Modelling of Stochastic Demand for Manufacturing Employment. VI, 158 pages. 1990.

Vol. 350: W.-B. Zhang, Economic Dynamics. X, 232 pages. 1990.

Vol. 351: A. Lewandowski, V. Volkovich (Eds.), Multiobjective Problems of Mathematical Programming. Proceedings, 1988. VII, 315 pages. 1991.

Vol. 352: O. van Hilten, Optimal Firm Behaviour in the Context of Technological Progress and a Business Cycle. XII, 229 pages. 1991.

Vol. 353: G. Ricci (Ed.), Decision Processes in Economics. Proceedings, 1989. III, 209 pages 1991.

Vol. 354: M. Ivaldi, A Structural Analysis of Expectation Formation. XII, 230 pages. 1991.

Vol. 355: M. Salomon. Deterministic Lotsizing Models for Production Planning. VII, 158 pages. 1991.

Vol. 356: P. Korhonen, A. Lewandowski, J . Wallenius (Eds.), Multiple Criteria Decision Support. Proceedings, 1989. XII, 393 pages. 1991.

Vol. 357: P. Zörnig, Degeneracy Graphs and Simplex Cycling. XV, 194 pages. 1991.

Vol. 358: P. Knottnerus, Linear Models with Correlated Disturbances. VIII, 196 pages. 1991.

Vol. 359: E. de Jong, Exchange Rate Determination and Optimal Economic Policy Under Various Exchange Rate Regimes. VII, 270 pages. 1991.

Vol. 360: P. Stalder, Regime Translations, Spillovers and Buffer Stocks. VI, 193 pages . 1991.

Vol. 361: C. F. Daganzo, Logistics Systems Analysis. X, 321 pages. 1991.

Vol. 362: F. Gehrels, Essays in Macroeconomics of an Open Economy. VII, 183 pages. 1991.

Vol. 363: C. Puppe, Distorted Probabilities and Choice under Risk. VIII, 100 pages . 1991

Vol. 364: B. Horvath, Are Policy Variables Exogenous? XII, 162 pages. 1991.

Vol. 365: G. A. Heuer, U. Leopold-Wildburger. Balanced Silverman Games on General Discrete Sets. V, 140 pages. 1991.

Vol. 366: J. Gruber (Ed.), Econometric Decision Models. Proceedings, 1989. VIII, 636 pages. 1991.

Vol. 367: M. Grauer, D. B. Pressmar (Eds.), Parallel Computing and Mathematical Optimization. Proceedings. V, 208 pages. 1991.

Vol. 368: M. Fedrizzi, J. Kacprzyk, M. Roubens (Eds.), Interactive Fuzzy Optimization. VII, 216 pages. 1991.

Vol. 369: R. Koblo, The Visible Hand. VIII, 131 pages. 1991.

370: M. J. Beckmann, M. N. Gopalan, R. Subramanian .), Stochastic Processes and their Applications. eedings, 1990. XLI, 292 pages. 1991.

ol. 371: A. Schmutzler, Flexibility and Adjustment to Information in Sequential Decision Problems. VIII, 198 pages. 1991.

Vol. 372: J. Esteban, The Social Viability of Money. X, 202 pages. 1991.

Vol. 373: A. Billot, Economic Theory of Fuzzy Equilibria. XIII, 164 pages. 1992.

Vol. 374: G. Pflug, U. Dieter (Eds.), Simulation and Optimization. Proceedings, 1990. X, 162 pages. 1992.

Vol. 375: S.-J. Chen, Ch.-L. Hwang, Fuzzy Multiple Attribute Decision Making. XII, 536 pages. 1992.

Vol. 376: K.-H. Jöckel, G. Rothe, W. Sendler (Eds.), Bootstrapping and Related Techniques. Proceedings, 1990. VIII, 247 pages. 1992.

Vol. 377: A. Villar, Operator Theorems with Applications to Distributive Problems and Equilibrium Models. XVI, 160 pages. 1992.

Vol. 378: W. Krabs, J. Zowe (Eds.), Modern Methods of Optimization. Proceedings, 1990. VIII, 348 pages. 1992.

Vol. 379: K. Marti (Ed.), Stochastic Optimization. Proceedings, 1990. VII, 182 pages. 1992.

Vol. 380: J. Odelstad, Invariance and Structural Dependence. XII, 245 pages. 1992.

Vol. 381: C. Giannini, Topics in Structural VAR Econometrics. XI, 131 pages. 1992.

Vol. 382: W. Oettli, D. Pallaschke (Eds.), Advances in Optimization. Proceedings, 1991. X, 527 pages. 1992.

Vol. 383: J. Vartiainen, Capital Accumulation in a Corporatist Economy. VII, 177 pages. 1992.

Vol. 384: A. Martina, Lectures on the Economic Theory of Taxation. XII, 313 pages. 1992.

Vol. 385: J. Gardeazabal, M. Regúlez, The Monetary Model of Exchange Rates and Cointegration. X, 194 pages. 1992.

Vol. 386: M. Desrochers, J.-M. Rousseau (Eds.), Computer-Aided Transit Scheduling. Proceedings, 1990. XIII, 432 pages. 1992.

Vol. 387: W. Gaertner, M. Klemisch-Ahlert, Social Choice and Bargaining Perspectives on Distributive Justice. VIII, 131 pages. 1992.

Vol. 388: D. Bartmann, M. J. Beckmann, Inventory Control. XV, 252 pages. 1992.

Vol. 389: B. Dutta, D. Mookherjee, T. Parthasarathy, T. Raghavan, D. Ray, S. Tijs (Eds.), Game Theory and Economic Applications. Proceedings, 1990. IX, 454 pages. 1992.

Vol. 390: G. Sorger, Minimum Impatience Theorem for Recursive Economic Models. X, 162 pages. 1992.

Vol. 391: C. Keser, Experimental Duopoly Markets with Demand Inertia. X, 150 pages. 1992.

Vol. 392: K. Frauendorfer, Stochastic Two-Stage Programming. VIII, 228 pages. 1992.

Vol. 393: B. Lucke, Price Stabilization on World Agricultural Markets. XI, 274 pages. 1992.

Vol. 394: Y.-J. Lai, C.-L. Hwang, Fuzzy Mathematical Programming. XIII, 301 pages. 1992.

Vol. 395: G. Haag, U. Mueller, K. G. Troitzsch (Eds.), Economic Evolution and Demographic Change. XVI, 409 pages. 1992.

Vol. 396: R. V. V. Vidal (Ed.), Applied Simulated Annealing. VIII, 358 pages. 1992.

Vol. 397: J. Wessels, A. P. Wierzbicki (Eds.), User-Oriented Methodology and Techniques of Decision Analysis and Support. Proceedings, 1991. XII, 295 pages. 1993.

Vol. 398: J.-P. Urbain, Exogeneity in Error Correction Models. XI, 189 pages. 1993.

Vol. 399: F. Gori, L. Geronazzo, M. Galeotti (Eds.), Nonlinear Dynamics in Economics and Social Sciences. Proceedings, 1991. VIII, 367 pages. 1993.

Vol. 400: H. Tanizaki, Nonlinear Filters. XII, 203 pages. 1993.

Vol. 401: K. Mosler, M. Scarsini, Stochastic Orders and Applications. V, 379 pages. 1993.

Vol. 402: A. van den Elzen, Adjustment Processes for Exchange Economies and Noncooperative Games. VII, 146 pages. 1993.

Vol. 403: G. Brennscheidt, Predictive Behavior. VI, 227 pages. 1993.

Vol. 404: Y.-J. Lai, Ch.-L. Hwang, Fuzzy Multiple Objective Decision Making. XIV, 475 pages. 1994.

Vol. 405: S. Komlósi, T. Rapcsák, S. Schaible (Eds.), Generalized Convexity. Proceedings, 1992. VIII, 404 pages. 1994.

Vol. 406: N. M. Hung, N. V. Quyen, Dynamic Timing Decisions Under Uncertainty. X, 194 pages. 1994.

Vol. 407: M. Ooms, Empirical Vector Autoregressive Modeling. XIII, 380 pages. 1994.

Vol. 408: K. Haase, Lotsizing and Scheduling for Production Planning. VIII, 118 pages. 1994.

Vol. 409: A. Sprecher, Resource-Constrained Project Scheduling. XII, 142 pages. 1994.

Vol. 410: R. Winkelmann, Count Data Models. XI, 213 pages. 1994.

Vol. 411: S. Dauzère-Péres, J.-B. Lasserre, An Integrated Approach in Production Planning and Scheduling. XVI, 137 pages. 1994.

Vol. 412: B. Kuon, Two-Person Bargaining Experiments with Incomplete Information. IX, 293 pages. 1994.

Vol. 413: R. Fiorito (Ed.), Inventory, Business Cycles and Monetary Transmission. VI, 287 pages. 1994.

Vol. 414: Y. Crama, A. Oerlemans, F. Spieksma, Production Planning in Automated Manufacturing. X, 210 pages. 1994.

Vol. 415: P. C. Nicola, Imperfect General Equilibrium. XI, 167 pages. 1994.

Vol. 416: H. S. J. Cesar, Control and Game Models of the Greenhouse Effect. XI, 225 pages. 1994.

Vol. 417: B. Ran, D. E. Boyce, Dynamic Urban Transportation Network Models. XV, 391 pages. 1994.

Vol. 418: P. Bogetoft, Non-Cooperative Planning Theory. XI, 309 pages. 1994.

Vol. 419: T. Maruyama, W. Takahashi (Eds.), Nonlinear and Convex Analysis in Economic Theory. VIII, 306 pages. 1995.

Vol. 420: M. Peeters. Time-To-Build. Interrelated Investment and Labour Demand Modelling. With Applications to Six OECD Countries. IX, 204 pages. 1995.

Vol. 421: C. Dang, Triangulations and Simplicial Methods. IX, 196 pages. 1995.

Vol. 422: D. S. Bridges, G. B. Mehta. Representations of Preference Orderings. X, 165 pages. 1995.

Vol. 423: K. Marti, P. Kall (Eds.), Stochastic Programming. Numerical Techniques and Engineering Applications. VIII, 351 pages. 1995.

Vol. 424: G. A. Heuer, U. Leopold-Wildburger, Silverman's Game. X, 283 pages. 1995.

Vol. 425: J. Kohlas. P.-A. Monney, A Mathematical Theory of Hints. XIII, 419 pages, 1995.

Vol. 426: B. Finkenstädt, Nonlinear Dynamics in Economics. IX, 156 pages. 1995.

Vol. 427: F. W. van Tongeren, Microsimulation Modelling of the Corporate Firm. XVII, 275 pages. 1995.

Vol. 428: A. A. Powell, Ch. W. Murphy, Inside a Modern Macroeconometric Model. XVIII, 424 pages. 1995.

Vol. 429: R. Durier, C. Michelot. Recent Developments in Optimization. VIII, 356 pages. 1995.

Vol. 430: J. R. Daduna, I. Branco, J. M. Pinto Paixão (Eds.), Computer-Aided Transit Scheduling. XIV, 374 pages. 1995.

Vol. 431: A. Aulin, Causal and Stochastic Elements in Business Cycles. XI, 116 pages. 1996.

Vol. 432: M. Tamiz (Ed.), Multi-Objective Programming and Goal Programming. VI, 359 pages. 1996.

Vol. 433: J. Menon, Exchange Rates and Prices. XIV, 313 pages. 1996.

Vol. 434: M. W. J. Blok, Dynamic Models of the Firm. VII, 193 pages. 1996.

Vol. 435: L. Chen, Interest Rate Dynamics, Derivatives Pricing, and Risk Management. XII, 149 pages. 1996.

Vol. 436: M. Klemisch-Ahlert, Bargaining in Economic and Ethical Environments. IX, 155 pages. 1996.

Vol. 437: C. Jordan, Batching and Scheduling. IX, 178 pages. 1996.

Vol. 438: A. Villar. General Equilibrium with Increasing Returns. XIII, 164 pages. 1996.

Vol. 439: M. Zenner, Learning to Become Rational. VII, 201 pages. 1996.

Vol. 440: W. Ryll, Litigation and Settlement in a Game with Incomplete Information. VIII, 174 pages. 1996.

Vol. 441: H. Dawid, Adaptive Learning by Genetic Algorithms. IX, 166 pages.1996.

Vol. 442: L. Corchón. Theories of Imperfectly Competitive Markets. XIII, 163 pages. 1996.

Vol. 443: G. Lang, On Overlapping Generations Models with Productive Capital. X, 98 pages. 1996.

Vol. 444: S. Jørgensen, G. Zaccour (Eds.), Dynamic Competitive Analysis in Marketing. X, 285 pages. 1996.

Vol. 445: A. H. Christer, S. Osaki, L. C. Thomas (Eds.), Stochastic Modelling in Innovative Manufactoring. X, 361 pages. 1997.

Vol. 446: G. Dhaene. Encompassing. X, 160 pages. 1

Vol. 447: A. Artale, Rings in Auctions. X, 172 pages. 1

Vol. 448: G. Fandel. T. Gal (Eds.). Multiple Criteria Decisi, Making. XII, 678 pages. 1997.

Vol. 449: F. Fang, M. Sanglier (Eds.). Complexity and Self-Organization in Social and Economic Systems. IX, 317 pages, 1997.

Vol. 450: P. M. Pardalos, D. W. Hearn, W. W. Hager. (Eds.), Network Optimization. VIII, 485 pages, 1997.

Vol. 451: M. Salge, Rational Bubbles. Theoretical Basis, Economic Relevance, and Empirical Evidence with a Special Emphasis on the German Stock Market.IX, 265 pages. 1997.

Vol. 452: P. Gritzmann, R. Horst. E. Sachs, R. Tichatschke (Eds.), Recent Advances in Optimization. VIII, 379 pages, 1997.

Vol. 453: A. S. Tangian, J. Gruber (Eds.), Constructing Scalar-Valued Objective Functions. VIII, 298 pages. 1997.

Vol. 454: H.-M. Krolzig. Markov-Switching Vector Autoregressions. XIV, 358 pages. 1997.

Vol. 455: R. Caballero, F. Ruiz. R. E. Steuer (Eds.). Advances in Multiple Objective and Goal Programming. VIII, 391 pages. 1997.

Vol. 456: R. Conte. R. Hegselmann, P. Terna (Eds.). Simulating Social Phenomena. VIII. 536 pages. 1997.

Vol. 457: C. Hsu. Volume and the Nonlinear Dynamics of Stock Returns. VIII, 133 pages. 1998.

Vol. 458: K. Marti. P. Kall (Eds.). Stochastic Programming Methods and Technical Applications. X, 437 pages. 1998.

Vol. 459: H. K. Ryu. D. J. Slottje. Measuring Trends in U.S. Income Inequality. XI, 195 pages. 1998.

Vol. 460: B. Fleischmann. J. A. E. E. van Nunen, M. G. Speranza, P. Stähly. Advances in Distribution Logistic. XI, 535 pages. 1998.

Vol. 461: U. Schmidt, Axiomatic Utility Theory under Risk. XV, 201 pages. 1998.

Vol. 462: L. von Auer, Dynamic Preferences. Choise Mechanisms, and Welfare. XII. 226 pages. 1998.

Vol. 463: G. Abraham-Frois (Ed.). Non-Linear Dynamics and Endogenous Cycles. VI. 204 pages. 1998.

Vol. 464: A. Aulin. The Impact of Science on Economic Growth and its Cycles. IX, 204 pages. 1998.

Vol. 465: T. J. Stewart, R. C. van den Honert (Eds.). Trends in Multicriteria Decision Making. X, 448 pages. 1998.

Vol. 466: A. Sadrieh, The Alternating Double Auction Market. VII, 350 pages. 1998.

Vol. 467: H. Hennig-Schmidt. Gargaining in a Video Experiment. Determinants of Boundedly Rational Behavior. XII, 221 pages. 1999.

Vol. 468: A. Ziegler, A Game Theory Analysis of Options. XIV, 145 pages. 1999.

Vol. 469: M. P. Vogel. Environmental Kuznets Curves. XIII, 197 pages. 1999.